"We Freeze to Please"

A History of NASA's Icing Research Tunnel and the Quest for Flight Safety

William M. Leary

The formation of ice on wings and other control surfaces of airplanes is one of the oldest and most vexing problems that aircraft engineers and scientists continue to face. While no easy, comprehensive answers exist, the staff at NASA's Icing Research Tunnel (IRT) at the Glenn Research Center in Cleveland has done pioneering work to make flight safer for experimental, commercial, and military customers.

The National Advisory Committee for Aeronautics (NACA) initiated government research on aircraft icing in the 1930s at its Langley facility in Virginia. Icing research shifted to the NACA's Cleveland facility in the 1940s. Initially there was little focus on icing at either location, as these facilities were more concerned with aerodynamics and engine development. With several high-profile fatal crashes of air mail carriers, however, the NACA soon realized the need for a leading research facility devoted to icing prevention and removal. The IRT began operation in 1944 and, despite renovations and periodic attempts to shut it down, has continued to function productively for almost 60 years.

In part because icing has proved so problematic over time, IRT researchers have been unusually open-minded in experimenting with a wide variety of substances, devices, and techniques. Early icing prevention experiments involved grease, pumping hot engine exhaust onto the wings, glycerin soap, mechanical and inflatable "boots," and even corn syrup. The IRT staff also looked abroad for ideas and later tried a German and Soviet technique of electromagnetism, to no avail. More recently, European polymer fluids have been more promising. The IRT even periodically had "amateur nights" in which a dentist's coating for children's teeth proved unequal to the demands of super-cooled water droplets blown at 100 miles per hour.

Despite many research dead-ends, IRT researchers have achieved great success over the years. They have developed important computer models, such as the LEWICE software, and made significant contributions to prevent ice buildup on turbine-powered commercial aircraft, helicopters, and military planes.

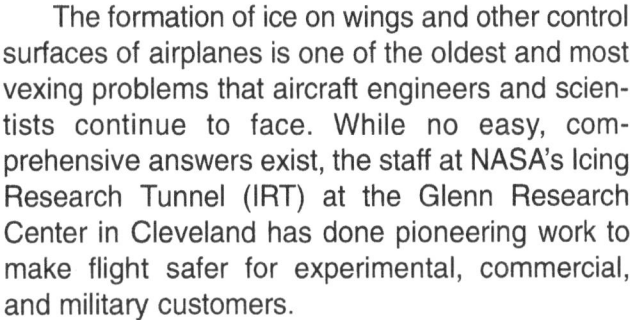

NASA SP-2002-4226

"We Freeze to Please"

A History of NASA's Icing Research Tunnel

and the Quest for Flight Safety

William M. Leary

The NASA History Series

National Aeronautics and Space Administration
Office of External Relations
NASA History Office
Washington, DC
2002

Contents

FOREWORD

In 1978, something wholly unexpected happened at what was then the NASA Lewis (now Glenn) Research Center—a "new" aircraft icing program was started. The story goes something like this: Milt Beheim, chief of the Wind Tunnel and Flight Division, was put in charge of the nearly defunct Icing Research Tunnel (IRT). It was in a state of serious disrepair, and no NASA program needed it. Industry used it sporadically. A propulsion and power laboratory, Lewis Research Center (LeRC) no longer had much interest in aircraft icing. Many influential people at LeRC were veterans of the successful NACA icing program of the 1940s and 1950s, and thus thought they had solved the icing problem a long time ago.

So what should Milt do with the IRT? In typical Beheim fashion, he researched aircraft icing, studied reports, and contacted U.S. aerospace companies and other civil and military government agencies. With encouragement and help from those contacts, he organized an International Workshop in Aircraft Icing in July 1978 at LeRC.

As you will read in this book, the workshop fully endorsed NASA's getting back into aircraft icing. Times had changed, the participants said. Growing commuter markets insisted on dependable flight schedules; an expanding private aircraft fleet wanted "all weather" capability for their expensive investment; military and civilian helicopters needed rotor-blade ice protection; the large transport aircraft sought more energy-efficient systems; and it was time to apply modern computers and instrumentation to the icing problem. Milt became a believer, and, to the surprise of many old "icingologists," he won the day by gaining approval to form a new Aircraft Icing Section in his division. Through the competitive process, I was selected to head this new section.

Over the next couple of years, we spent many hours visiting the offices at NASA Headquarters seeking funding for the new section. When the opportunity later arose, Milt organized a successful effort to finance the first major upgrade of the IRT. With that upgrade, the IRT went from LeRC's most humble facility to its most modern wind tunnel, a prototype for future upgrades to its other tunnels.

That first IRT upgrade could not have been more timely. On 13 January 1982, the Air Florida flight 90 accident at Washington National Airport alerted the nation to the lethal hazards of ground icing and influenced NASA's decision to approve the first upgrade to the IRT. Shortly after the upgrade was complete, the Boeing Airplane Co., the FAA, the Association of European Airlines, and the fluids manufacturers approached us about conducting an extensive joint test program in the IRT to evaluate the effects of ground de-icing/anti-icing fluids on takeoff aerodynamics. We eagerly agreed to this crucial international safety program. We knew that the upgraded IRT, with its modern control systems, could do the job. The IRT and its test crews performed flawlessly, prompting the Boeing engineers to remark that the IRT productivity was at least as high as their best performing wind tunnels. The IRT test results formed the basis for new ground de-icing regulations, which the FAA promptly promulgated to the air transportation industry. That winter, following the IRT tests, the airline industry was using the newly tested and approved fluids.

The aerospace community came to regard the IRT as a unique national resource. In 1987, the American Society of Mechanical Engineers designated the IRT an "International Historic Mechanical Engineering Landmark" for its leading role in making aviation safer for everyone. When I retired from NASA in 1994, the IRT had long been one of NASA's most heavily used windtunnels. Its annual test time of about 1,000 hours was divided nearly equally between government research and industrial development. In 2002, nearly 24 years after its "new" start, the IRT is now NASA's busiest wind tunnel.

The new icing program began with three NASA-funded study contracts, in which the U.S. aerospace industry told us what it needed. We, in turn, set up a program to address all of those needs. Its key objectives were as follows:

1. Develop computer codes that would: a) predict water droplet collection on aircraft surfaces, b) model the ice buildup on aircraft surfaces, and c) provide design tools for various ice-protection systems.
2. Assess and, where appropriate, experimentally evaluate some recently proposed ice-protection systems, such as microwave systems and electro-impulse systems. Fund the development of new ice-protection concepts through the proof of principle stage.
3. Assess and, where appropriate, experimentally evaluate new icing instruments for the detection of ice on aircraft and the measurement of super-cooled cloud properties in flight. Fund the development of new icing instruments and their calibration procedures through the proof of principle stage.
4. Upgrade the IRT to provide improved control of the air speed, air temperature, and super-cooled cloud properties.
5. Conduct aircraft flight tests to assess the effects of ice accretion on aircraft performance, to ensure that IRT test results accurately simulate natural icing, and to extend the existing database on the physical properties of super-cooled icing clouds. The aircraft should be fully instrumented for aircraft performance measurements and for cloud property measurements.

Although the three study contracts formed the foundation of our icing program, it evolved further through frequent NASA-industry workshops and peer reviews. Industry or other government agencies often proposed programs to address urgent technology or safety needs. With such a diverse and extensive list of objectives, it was clear that NASA could not singlehandedly support this large effort in either funding or staff. So we did it by coordinating personnel, test equipment, and experimental facilities within the U.S. government, industry, universities, and, on occasion, within other countries. We wrote interagency agreements with the FAA, Army, Navy, and Air Force. We also forged international agreements in icing with Canada, France, and the United Kingdom. Some of these interagency agreements were accompanied by fund transfers to NASA, and these were an important part of our total funding picture. Another important source of funds was the SBIR (small business innovative research) program at LeRC, through which nearly all of our advanced ice-protection concepts were funded.

A good example of coordination that required both national and international resources was the program to evaluate pneumatic boots for helicopter rotors. The BFGoodrich Company supplied pneumatic boots and system hardware for IRT model tests and helicopter flight tests, and participated in all tests. The U.S. Army supplied a UH-1H helicopter, its pilots, and its support crews for all flight tests. The Bell Helicopter Company supplied experienced test engineers to assess any safety hazards and to follow all tests. Canada supplied its Hover Icing Spray Rig. The U.S. Army supplied their in-flight icing simulator Helicopter Icing Spray System (HISS), a custom CH-47 heavy-lift helicopter fitted with water tanks and spray bars, and all the support aircraft and crews. NASA helped calibrate the HISS by flying its heavily instrumented Twin Otter in the super-cooled cloud behind the HISS. The program began with NASA testing several boot configurations in the IRT to determine the best tube arrangement for effective ice removal and minimal drag increase. Then Army personnel flew the UH-1H in clear air to evaluate the effects of the inflated boots on helicopter handling qualities. Next they flew hover tests in Canada's Hover Icing Spray Rig. Then they flew behind the HISS at Edwards Air Force Base in California. Again they flew in natural icing in Minnesota. Finally, the Army tested the boots on a UH-1H in its special rain and sand erosion facility at Fort Rucker, Alabama.

By coordinating and sharing our resources, we became a close-knit icing community in which good communications flourished. Industry and government agencies got test results promptly, because they participated in the tests. Master's and doctoral students worked closely with industry to ensure their NASA grant research was relevant and that our new NASA computer codes gained wide distribution.

With our "new" program came the heavy development of computer codes. These included droplet trajectory codes, ice-accretion modeling codes, and ice-protection system design codes. These highly useful codes have been distributed widely throughout industry and government. Hundreds of engineers use the LEWICE ice-accretion modeling code, and the FAA now accepts it as part of their icing certification process. NASA continues to improve these codes and verify them against experimental data. Because of our close ties with the European icing community, they too are familiar with LEWICE and our other codes—a favorable situation for those U.S. airplane manufacturers who must certify their products in Europe.

Although we developed several new ice-protection and ice-detection concepts early on, mainly through SBIR contracts, industry did not adopt any initially. But after I retired and Tom Bond became the branch chief, his team succeeded, via an SBIR contract, in getting a new aircraft ice-protection system approved by the FAA in 2001. It was the first new ice-protection system approved by the FAA in 40 years! The system, built by Cox & Company, Inc., New York, NY, is a hybrid that uses both thermal anti-icing and electro-mechanical expulsion de-icing. The system is adaptable and uses much less energy than other systems that provide equivalent protection. The system is in production for Raytheon Aircraft's Premier I business jet.

Another internationally recognized contribution of Tom Bond's branch is pilot education for improved flight safety in icing conditions. NASA has made three pilot training videos and one training CD (compact disk). These widely distributed videos and CDs address the icing environment, flight preparedness and strategies for avoiding icing conditions, stall due to tailplane ice, and loss of control due to wing ice. According to Tom Bond, "The tail stall recovery techniques developed in the Tailplane Icing Program have resulted in averted accidents and saved lives. NASA has received numerous accounts of pilots who have used these operational procedures and successfully prevented a loss of control due to a tail stall."

For 10 years, Tom Ratvasky has been the technical program manager and lead flight research engineer on NASA's icing flight research aircraft. The American Institute of Aeronautics and Astronautics recognized Mr. Ratvasky's leadership in conceiving and producing these instructional materials by awarding him the prestigious 2002 Losey Atmospheric Science Award.

I believe that we can say with pride that the "new" aircraft icing program at Glenn Research Center has continued the tradition of the "old" program by contributing valuable technological and safety support to the aircraft industry.

One final note: the title of this book, "We Freeze to Please," was the brainchild of Mr. Al Dalgleish, a branch chief in the Test Installations Division. He used to answer the telephone by saying "Icing Research Tunnel, where we freeze to please." Besides being a play on words, it was Al's way of saying "we will cooperate with you to get your job done." The "We Freeze to Please" logo made its way onto the IRT baseball caps. Nearly everyone who tested in the IRT went home with an IRT baseball cap. Visitors from other countries took them home. The logo sent the message far and wide that the IRT staff was there to help. As you read this book, you will surely be struck by the pride and esprit of the IRT operators, operations engineers, and research engineers. This hardworking and productive staff made my association with icing a pleasure. I thank you one and all.

-John J. (Jack) Reinmann
Cleveland, Ohio
April 2002

Preface

Icing research has received only limited attention from historians. Yet the problem of icing has challenged aviators and aircraft manufacturers since the earliest days of powered flight—and continues to do so. At various times in the past, victory has been declared over the menace of icing, but these claims have always proved premature. Despite more than seven decades of research into the phenomena, much remains to be learned about the nature of icing and how best to respond to it.

In the United States, the National Advisory Committee for Aeronautics (NACA) and its successor, the National Aeronautics and Space Administration (NASA), have led the way in investigating the interaction between aircraft and the icing environment, as well as in developing various means to protect fixed- and rotary-wing machines. To be sure, icing research has never been given a high priority. When the NACA began to investigate icing during the 1930s, most of the engineers at the Langley Memorial Aeronautical Laboratory were far more interested in advances in aerodynamics than they were in icing. When work shifted to the new Aircraft Engine Research Laboratory in Cleveland in the 1940s, icing research represented only a minor interest of a laboratory that was devoted to engine development. Later, the demands of the space age overshadowed NASA's work on aeronautics. Within this limited context, icing investigations usually ranked low on the aeronautical research agenda.

Perhaps because of their lack of status in the NACA/NASA world, icing researchers tended to form a close-knit group. Untroubled by distinctions between fundamental research and practical engineering, they believed that they were doing important work and were making a significant contribution to safety. They found other icing enthusiasts in industry, academia, and government agencies, both in the United States and abroad, and eventually came together in an international "icing community."

Esprit de corps among individuals concerned with icing problems tended to be high. Over and over again during the course of my research for this study, interviewees would tell me that their association with icing research was the highpoint of their careers. Their attitude was infectious, making my work not only a learning experience, but also an enjoyable one.

In telling the story of NACA/NASA icing research, I have focused on the role of the Icing Research Tunnel (IRT) at the Cleveland laboratory. Because the experiments that were conducted in the IRT formed only part of a broader investigation into icing which encompassed flight research and computer simulation, I have included information about these topics while keeping the tunnel at the center of my study. Also, I have attempted to place the work of the NACA/NASA in the broader context of the icing problems faced by the international aviation community.

I have been assisted by many people along the way. Kenneth E. Zaremba, now management analyst at the Glenn Research Center, first had the idea for a book on icing research and persuaded NASA Headquarters to support the project. Susan L. Kevdzija, IRT facility manager and enthusiastic supporter from the beginning, hosted my first visit to Glenn and opened many doors for me. William P. Sexton conducted a tour of the icing tunnel and shared with me his deep affection for the facility. Robert F. Ide kindly allowed me to observe the operation of the tunnel during a calibration run.

Kevin P. Coleman, history coordinator at Glenn, looked after my care and feeding during visits to the laboratory, and I am grateful to him. Laura M. Bagnell and Mary E. Carson of the Imaging Technology Center opened to me the rich photographic collection at Glenn. During a visit to the NASA History Office in Washington, Jane H. Odom facilitated my access to icing records and made my brief stay in the office a most pleasant one. Also at Headquarters, NASA chief historian Roger D. Launius extended his full support to the project, and I am grateful to him.

Edmund Preston, historian at the Federal Aviation Administration (FAA) and an old friend, always promptly responded to my requests for FAA material. Another old friend, Michael H. Gorn, took time during a personal tragedy to send me a draft of his now published study of flight research at the NACA and NASA. H. Garland Gouger and Erik Conway made my visit to the archives at the Langley Research Center a most rewarding experience.

Bonita S. Smith, InDyne, Inc., employee and archivist/historian at the Glenn Center, has been my strong right arm throughout the research for this book. An accomplished archivist, she has searched for documents, located obscure technical reports, arranged interviews, located retired employees, and cheerfully responded to a thousand and one requests. Without her assistance, this study would have taken far longer—and might never have been finished. I cannot thank her enough.

Thanks are also due to the professionals at NASA Headquarters who made this book physically possible. In the NASA History Office, Louise Alstork edited the manuscript and prepared the index. In the Printing and Design Office, Joel Vendette and Steve Oberti expertly laid out the book, Michelle Cheston carefully edited it, and David Dixon handled the printing. My sincere appreciation goes out to all these people.

The many individuals who submitted to interviews, wrote letters and e-mails, provided material from their personal files, and generally contributed substantially to making their study possible are listed in the Essay on Sources. While I am grateful to all of them, I should note that John J. Reinmann took a special interest in the project and spent a great deal of time attempting to educate me on the nature of icing research. As I managed to avoid physics and chemistry in high school, this was no small task.

In writing this book, my fondest hope is that I have done justice to the efforts of the many men and women who have devoted a substantial portion of their lives to defeating the menace of icing and making the skies safer for us all.

William M. Leary
Athens, Georgia

Chapter 1
The Beginning of Icing Research

The dangers posed by icing rarely concerned aviators during the first two decades of powered flight. Lacking the instruments necessary to fly without visual references, pilots did their best to avoid clouds. As a result, encounters with icing seldom happened and were always inadvertent. The situation changed in the mid-1920s when the intrepid aviators of the U.S. Air Mail Service attempted to maintain scheduled day-and-night operations between New York and Chicago. These instrument-flying pioneers were the first group of flyers to face the icing menace on a regular basis. As one of their pilots noted at the time about the hazards of the New York-Chicago route, "the greatest of all our problems is ice."[1]

A typical encounter took place during the early morning hours of 23 December 1926. Pilot Warren Williams was en route from Cleveland to Chicago with 321 pounds of mail. He was flying underneath an overcast sky until low clouds at Woodville blocked his way. He decided to fly on top, as the cloud layer seemed only 1,000-feet thick. He went on instruments, monitoring his gyroscopic turn indicator, ball-bank indicator, and airspeed. As the Douglas M4 biplane began to climb, Williams felt his controls grow "mushy." His turn indicator malfunctioned; his compass began to spin; his altimeter unwound. Williams fought the controls, but without success. As the ground approached, he cut the throttle and jumped. He pulled the rip cord on his recently issued parachute and floated down safely from 300 feet.[2]

Williams was lucky to have survived his encounter with icing. Fellow pilot John F. Milatzo was not as fortunate. Shortly after midnight on 22 April 1927, while en route from Chicago to New York with the mail, Milatzo crashed into a field during a severe snow and sleet storm. He was killed.[3]

---

[1] Wesley L. Smith, "Weather Problems Peculiar to the New York-Chicago Airway," *Monthly Weather Review* 57 (December 1929): 503–06.

[2] William M. Leary, *Aerial Pioneers: The U.S. Air Mail Service, 1918–1927* (Washington, DC: Smithsonian Institution Press, 1985), p. 234.

[3] Ibid., pp. 234–35.

There were various attempts made to deal with the icing problem during the 1920s. The U.S. Air Mail Service, for example, worked closely with Army Air Service technicians at McCook Field, the major Air Service research facility in Dayton, Ohio, to find some answers. In 1925, the Army used a small wind tunnel that had been set up in a refrigerated room at McCook to study the formation of ice on pitot tubes. Although the Army failed to come up with a solution, instrument manufacturers later developed electrically heated pitot-static tubes.[4]

The National Advisory Committee for Aeronautics (NACA) first turned its attention to the icing problem in 1928, thanks to the initiative of George W. Lewis, Director of Aeronautical Research. On 10 February, Lewis wrote to Brig. Gen. William E. Gillmore, chief of the Air Material Division of the Air Corps, seeking information about the military's experience with icing. While Air Corps pilots had a number of encounters with ice, Gillmore responded on 28 February, a search of records produced only a single report. "No active research has yet been undertaken at this [Material] Division on the subject of ice formation," Gillmore continued. Nonetheless, he was prepared to offer his advice on how to deal with the problem. Heat could be applied to the parts of the airplane on which ice usually formed, he noted, although the means by which this could be accomplished remained untested. Also, he held out the hope that waterproof finishes might work, as they did for the aquatic birds that often passed through regions where ice frequently forms. "Propellers," he noted, "have been greased in some cases with apparent success in preventing ice formation."

Gillmore went on to suggest a variety of approaches that might be adopted to investigate the problem of icing. Flight research, he suggested, would probably be the most expensive alternative but also would be "the most certain to achieve results." Also, space could be secured in "a cold storage plant, where an air-conditioning apparatus could be installed to create the ice-forming atmosphere in which to whirl model airfoils." While several government agencies might become involved in seeking a solution to the icing problem, Gillmore concluded, the NACA should coordinate the various efforts. Accordingly, he recommended "that an authorization for research on this subject by the NACA be approved."[5]

No doubt as a result of Lewis's initiative, the Navy's Bureau of Aeronautics (BuAer) joined with the Air Corps in supporting a program of research into icing. "The problem

---

[4] Bradley Smith, "Icing Wings," *U.S. Air Services* 15 (April 1930): 22–25; Montgomery Knight and William C. Clay, "Refrigerated Wind Tunnel Tests on Surface Coatings for Preventing Ice Formation," NACA TN 339 (May 1930).

[5] Gillmore to Lewis, 28 February 1928, File RA 247, Historical Archive, Floyd C. Thompson Technical Library, Langley Research Center, Hampton, VA.

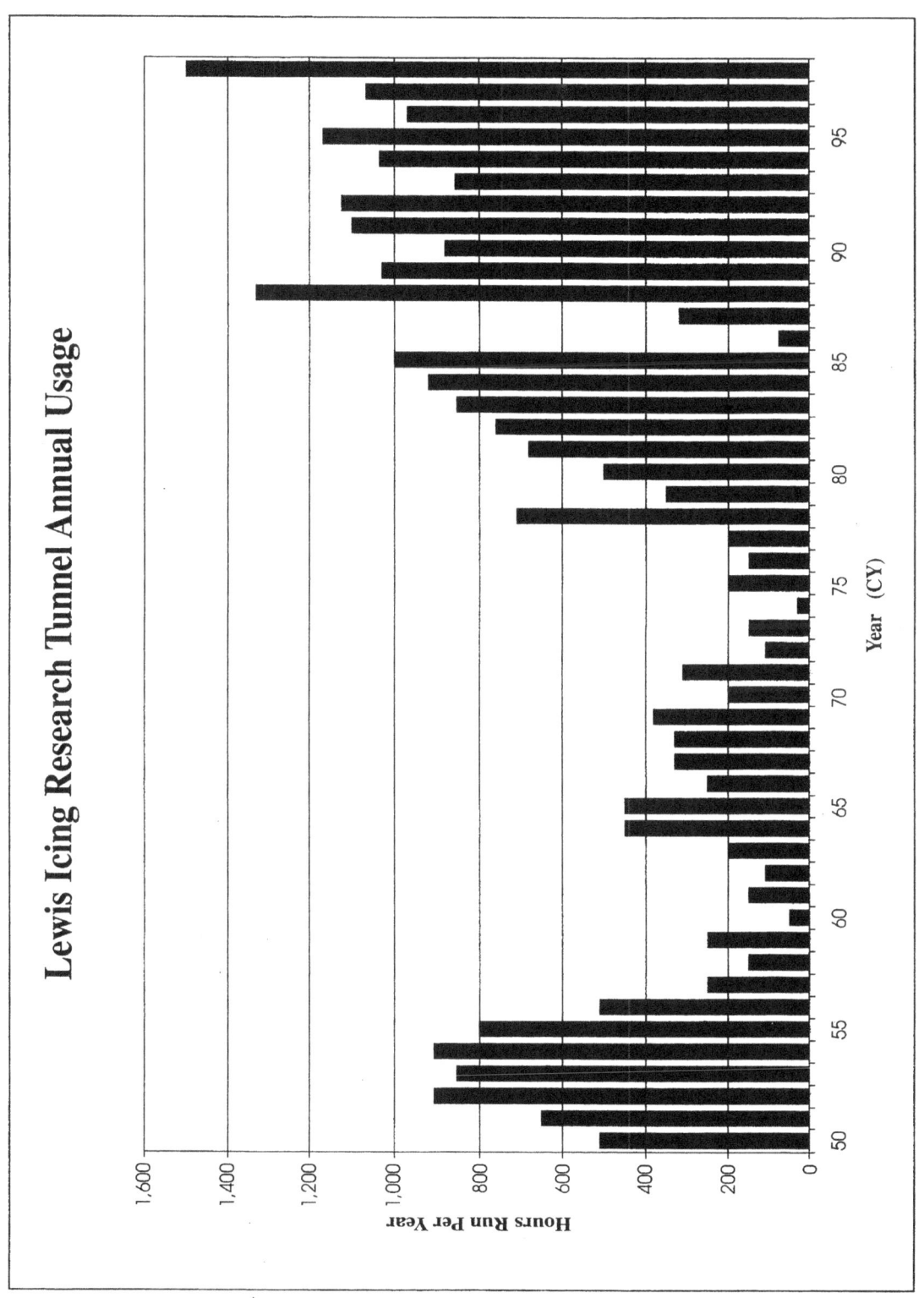

Figure 1–1. Lewis Icing Research Tunnel annual usage.

is one of great interest to Naval aeronautics," Rear Adm. William A. Moffett, chief of BuAer, wrote to Lewis on March 12, "and one in which very little data is available. A solution to this problem is considered to be one of great importance." Moffett went on to request that the NACA "undertake research on this problem with a view of determining the conditions under which ice forms on the structures of aircraft; second, possible preventative means; and third, the development of an instrument or instruments which would indicate to the pilot conditions of temperature and humidity under which ice formation takes place."[6]

One day later, citing Gillmore's request and anticipating the letter from BuAer, Lewis told the NACA's Langley Memorial Aeronautical Laboratory (LMAL) in Hampton, Virginia, to begin planning an investigation into the icing problem. He noted that the Department of Commerce, which also had an interest in the subject, had at first suggested that the Bureau of Standards erect a small wind tunnel in their altitude chamber and devise a system to control temperature and humidity in order to investigate conditions under which ice would form on metal and fabric surfaces. Upon further consideration, however, it had been decided to conduct the preliminary investigation at the Langley laboratory. "This investigation," Lewis continued, "could be made in the 6-inch wind tunnel, using a small metal airfoil section. The temperature of the airfoil section could be controlled by the expansion of CO_2 . . . and the humidity controlled either by water spray in the air stream or by admitting steam in the air stream. [If] the temperature of the air stream could also be controlled by the expansion of the CO_2, by lagging the small tunnel with Cellotex or cork, and by placing a box or chamber around the test section, fairly constant conditions could be obtained."

Lewis hoped that the investigation would produce information on the temperature and humidity at which ice will form on metal and fabric surfaces, develop an instrument to alert pilots to conditions under which icing could be expected, and suggest methods of preventing ice formation "such as heating the wings or coating the wing or surface with some material which will prevent deposition of water."[7]

Two weeks later, on 28 March, William P. MacCracken, Jr., assistant secretary of commerce for aeronautics, called a special conference of representatives of the Army Air Corps, Navy Bureau of Aeronautics, Weather Bureau, Bureau of Standards, and the NACA to study "the causes and prevention of ice formation on aircraft." Lewis was able to report at this meeting that the NACA had already begun such a study. Reports from pilots who had experienced icing conditions were being reviewed, and research flights into icing were underway. Also, the NACA had plans for tests "in a special wind tunnel

[6] Moffett to Lewis, 12 March 1928, File RA 247, Langley Library.

[7] Lewis to LMAL, 13 March 1928, RA 247, Langley Library.

in which atmospheric conditions will be simulated." The day after this meeting, Lewis informed Langley "that in as much as the National Advisory Committee for Aeronautics had undertaken this problem and was working on it, [it was agreed that] all activities along this line should be coordinated and should be reported to [the NACA]."[8]

The final step necessary to formalize the NACA's investigation into the icing problem came on 28 June 1928, when the organization's Executive Committee approved Research Authorization (RA) 247, "Ice Formation on Aircraft." RA 247 declared the purpose of the investigation was: "To determine the conditions under which ice forms on the structures of aircraft; to develop possible means of prevention; and to develop an instrument or instruments which will indicate to the pilot conditions of temperature and humidity under which ice formation takes place."[9]

Even before approval of the research authorization, NACA chief pilot Thomas Carroll had laid the groundwork for a flight investigation of icing. Carroll had been interested in the problem ever since Air Mail Service pilots had begun to report their difficulties with icing. "The most difficult aspect of the matter," he wrote to Henry J. E. Reid, engineer-in-charge at the Langley Laboratory, on 17 March 1928, "appears to lie in an almost total ignorance of the conditions which must exist to produce the phenomena and . . . the double difficulty of artificially reproducing the conditions." He recommended that the NACA's Flight Operations section conduct a research program into inflight icing on a priority basis. To accomplish the task, he wanted "a supercharged airplane" that would be equipped with automatically recording air temperature thermometers that also would be visible to the pilot. In addition, he sought "an automatically driven motion picture camera placed and focused at short range on certain struts and surfaces on the airplane to be controlled by the pilot or observer to provide additional data to visual observation." He proposed to seek out cloud formations with a range of temperatures that were conducive to icing. Together with investigating the phenomena of icing, Carroll also wanted to experiment with preventive measures, such as "heat control, oiling, etc."[10]

RA 247 meant that Carroll would get his opportunity to investigate icing. The NACA equipped a Vought VE-7 with small auxiliary surfaces and aerodynamic shapes that were similar to the struts, wires, pitot heads, and other areas that were vulnerable to icing. Research pilots Carroll and William H. McAvoy set out in search of cloud formations where ice was likely to be encountered. They recognized and described the different

---

[8] "Conference at Department of Commerce Regarding Formation of Ice on Aircraft," 28 March 1928; Lewis to LMAL, 29 March 1928; both in RA 247, Langley Library.

[9] Research Authorization 247, copy in RA 247, Langley Library.

[10] Carroll to H. J. E. Reid, 17 March 1928, RA 247, Langley Library.

types of ice that they found. Glaze ice, which formed at temperatures just below freezing, was clear and tended to protrude from the leading edge of airfoils. Rime ice, on the other hand, which occurred at lower temperatures and was opaque, usually took on a stream-line shape. Performance penalties, they noted, were caused more by distortions to the shape of the airfoil than by the weight of the ice. Engineers knew, of course, that a cubic foot of water weighed 62.5 pounds. When water expanded on freezing, the weight decreased to 56 pounds per cubic foot. Even a biplane, with its wires and struts, was unlikely to accrete more than seven cubic feet of ice, or some 400 pounds, during the most severe icing conditions. While the added weight would cause some performance penalties, it would not constitute a dangerous overload.

Aeronautical engineers also knew that air flowing over the upper surface of a wing produced lift. To ensure that nothing disturbed the even flow of air, the airfoil surface was gently rounded. When ice formed on this surface, especially on the leading edge of the airfoil, turbulence and eddy currents caused by the non-streamlined shape destroyed lift. At some point, the airfoil would stall, and the airplane would fall out of the sky.

Carroll and McAvoy pointed out that a number of means of preventing or removing the ice formations had been suggested. The most frequently tried method had been the use of oil or grease to reduce the adhesion of the ice to vulnerable parts of the airplane. To date, however, the application of oil and grease had failed to produce worthwhile results. In fact, it seemed that the use of these substances might even hasten the forma-tion of ice. Suggestions had also been made so that engine exhaust heat might be piped through the leading edge of the wing to melt the ice or prevent it from forming—a method that might be worth further investigation. But given the current state of ice pre-vention, Carroll and McAvoy concluded, "safety . . . obviously lies in avoidance."[11]

The NACA's laboratory research also began in 1928 when it placed into operation a refrigerated wind tunnel at Langley. The facility consisted of a metal shell that was insu-lated by layers of cork and wood. Air temperature inside the tunnel was lowered and regulated by brine, which was cooled by a commercial refrigerating apparatus that flowed through hollow metal guide vanes. A propeller circulated the air. Water droplets, their size regulated by air and water pressure, came through four spray nozzles. Double glass doors and windows permitted observation and photographs of models in the test chamber.

This first refrigerated icing research tunnel had two main problems. One was its small size—the air stream was only 6 inches in diameter. The other major difficulty related to the formation of small water droplets. "Considerable time," the NACA's

---

[11] Carroll and McAvoy, "The Formation of Ice Upon Exposed Parts of an Airplane," NACA TN 293 (1928), and "The Formation of Ice Upon Airplanes in Flight," TN 313 (1929). See also Bradley Jones, "Icy Wings," *U.S. Air Services* 15 (April 1930): 22–25.

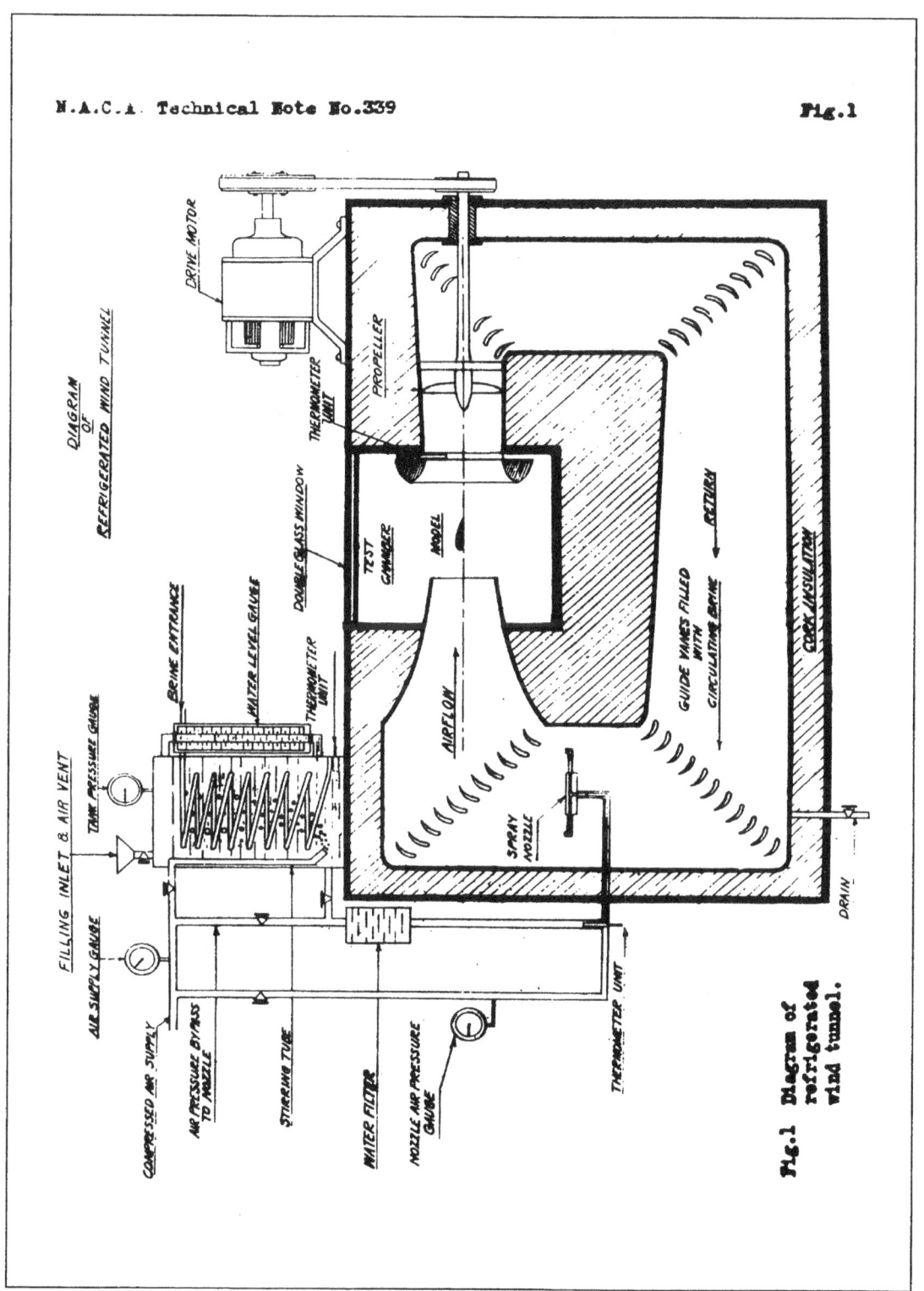

Figure 1–2. The NACA's first Refrigerated Icing Tunnel, 1930.

Annual Report for 1929 noted, had been devoted "to means for controlling the amount of water sprayed into the air stream, the size of water particles, and the temperatures of the air and water." Commercial spray nozzles simply could not produce the small water droplets that were found in natural icing. Nonetheless, it was possible to use the facility for some icing research; although fuller experimental capabilities would come only after the construction of an icing research tunnel in the 1940s.[12]

The first tests in the tunnel employed a section of a Clark Y mahogany airfoil. Widely used in the design of wings for aircraft during the 1920s, the high-lift airfoil featured a 3-inch chord and a 12-inch span. Researchers Montgomery Knight and William C. Clay left half the airfoil bare and brushed the other half with a thin coating of substances designed to retard the formation of ice. They first investigated six insoluble compounds: light and heavy lubricating oil, cup grease, Vaseline, paraffin, and simonize wax. None of the coatings prevented ice from accreting on the model. "The drops adhered to the surface," they reported, "especially at the stagnation point of the leading edge, and [they] froze quite readily as on the bare wing."

The researchers next tried five soluble substances: glycerin, glycerin and calcium chloride, molasses and calcium chloride, a hardened sugar solution, and a hardened glucose solution. The first three were brushed on the airfoil, but the sugar and glucose had to be boiled down and applied while hot. When they hardened, Knight and Clay noted, they had the consistency of "taffy candy." The solubles were supposed to dissolve with water as it struck the airfoil and lower the freezing point so that ice would not form. The glycerin and calcium chlorate solutions, however, immediately blew back from the leading edge and left it bare. The sugar and glucose solutions remained on the airfoil, but ice built up on top of them.

Knight and Clay also tested corn syrup, honey, glycerin soap, commercial paint, and goose grease. All proved disappointing except the White Karo corn syrup, which seemed to provide some protection against ice accretion and merited further study. Perhaps the most useful information derived from the first of what would prove a lengthy quest for ice-phobic materials was the observation that ice formed only on the leading edge of the airfoil in all tests. "Any preventive compound," they concluded, "need be applied only to that part of the wing to be effective."[13]

Flight tests brought more promising results. In June 1931, famed physicist Theodore Theodorsen and researcher William C. Clay reported on the use of engine

---

[12] *Fifteenth Annual Report of the National Advisory Committee for Aeronautics*, 1929 (Washington, DC: Government Printing Office, 1930), pp. 25–26.

[13] Knight and Clay, "Refrigerated Wind Tunnel Tests on Surface Coatings for Preventing Ice Formation," NACA TN 339 (1930).

exhaust heat as a means to prevent ice from forming. They mounted a modified Clark Y airfoil of approximately full-scale dimensions on a Fairchild F-17 monoplane and inserted a small boiler in the engine's exhaust pipe. Steam passed through a conducting pipe and entered the leading edge of the airfoil by means of a distribution pipe equipped with small holes. The researchers mounted spraying jets 4 feet in front of the airfoil. As the airplane was flown into temperatures as low as 18 °F, water was sprayed on the airfoil.

"The most essential result obtained in this study," Theodorsen and Clay wrote, "is the fact that ample heat is available in the exhaust and in the cooling water for the purpose of ice prevention." The successful design of an airplane immune from the dangers of ice accumulation, they confidently but erroneously predicted, was now "only a matter of technical development."[14]

Theodorsen and Clay also used the refrigerated wind tunnel at Langley for their icing research. After the NACA received reports about gasoline tank vents freezing in flight, the two researchers developed a program to test various configurations in the refrigerated wind tunnel. The primary function of the vents was to maintain the pressure inside the gasoline tank approximately equal to that of the atmosphere. The size, position, and location of the vent pipes differed with different types of aircraft, and the designs seemed "more or less at random."

The researchers positioned eleven different vent pipes in the test section of the refrigerated wind tunnel. The pipes varied in tube diameter from 0.125-inch to 0.5-inch and in shape from straight to L-shaped to U-shaped. They were subjected to the icing spray until the ends froze over and plugged the vent. Experiments showed that tubes perpendicular to the airstream consistently froze over, with the time required to plug the vent varying with the diameter of the tube. The opening of the tubes pointed aft, however, did not accrete ice.

As a result of their investigation, Theodorsen and Clay were able to make a specific recommendation to manufacturers and operators. A 0.75-inch tube, bent at a right angle and placed with the open end pointed downstream, they concluded, would be "the safest arrangement for gasoline tank vents . . . and also the most practical, with respect to gas tank pressure." Harry A. Sutton, chief engineer for American Airways, who earlier had reported a problem with the icing of gasoline tank vent pipes, expressed his appreciation to Dr. Lewis for the investigation and reported in October 1931 that "we are installing the type of vent recommended in your report."[15]

---

[14] NACA Report 403 (1933). Although published in 1933, this report was written on 12 June 1931.

[15] "The Prevention of Ice Formation on Gasoline Tank Vents," NACA TN 394; Sutton to Lewis, 16 March and 17 October 1931, RA 247, Langley Library.

While the NACA believed that the application of heat eventually would solve the problem of inflight icing, a more immediate solution seemed at hand during the early 1930s, thanks primarily to the work of William C. Geer. A retired scientist, Geer had graduated from Cornell University in 1905 with a doctorate in chemistry and joined the BFGoodrich Company of Akron, Ohio, in 1907 as their chief chemist. In 1927, two years after he had retired from Goodrich due to ill health, Geer became interested in the airplane icing problem. He knew that there had been sporadic research since 1922 on ice-phobic liquids and wing fabrics, but these early efforts had produced no satisfactory results. Geer decided to try his own experiments. He built a small research laboratory and began to test chemical methods to prevent the formation of ice.[16]

By 1929, Geer's work had showed sufficient promise to attract the attention of the Daniel Guggenheim Fund for the Promotion of Aeronautics. As part of its grant program to enhance aeronautical safety, the Guggenheim Fund gave Geer $10,000 to conduct further research. Working with Dr. Merit Scott of Union College, Geer arranged with the Department of Physics at Cornell to build a small icing research tunnel. The facility featured a 7-inch by 7-inch test section and a 3-inch circular throat, with the temperature lowered by ice.[17]

Tests in the tunnel suggested that oiled rubber sheets that covered the vulnerable parts of the airplane showed considerable promise. Coated with a mixture of 4 parts pine, 4 parts diethylthalate, and 1 part castor oil, the rubber sheet retarded the accumulation of ice. The major problem was to get rid of the ice that managed to form on the sheet. Working with B. F. Goodrich, Geer came up with an "expanding rubber sheet" or "ice-removing overshoe." The coated rubber sheet was placed on the leading edge of an airfoil in the tunnel, and air pressure was used to inflate the sheet and remove the ice.

Practical tests of the device were conducted in late March and April of 1930. Wesley Smith, a former Air Mail Service pilot who was now operations manager for National Air Transport, flew three test runs with the overshoe or boot. The test section consisted of laced-on overshoes that were 36 inches long by 15 inches wide. Two tubes, 2 inches in diameter, supplied air to inflate the boots. During the flights, Goodrich engineer Russell S. Colley sat on an orange crate in the mail compartment of the airplane and used a bicycle pump to deliver air into the tubes, alternating from one tube to the other by means of a manually operated valve. The boot worked well on a flight from Cleveland to

---

[16] William C. Geer, "The Ice Hazard on Airplanes," *Aeronautical Engineering* 4 (1932): 33–36.

[17] Ibid.; Richard P. Hallion, *Legacy of Flight: The Guggenheim Contribution to American Aviation* (Seattle: University of Washington Press, 1977), pp. 109–10.

Buffalo in heavy icing conditions. The main problem was ice forming on the unprotected propeller, which Smith removed by violent sideslips.[18]

Goodrich was so impressed with the flight tests that he decided to build a large icing research tunnel at Akron. By far the largest facility of its kind in the world, the tunnel measured 10 feet by 40 feet, with a test chamber 3 feet by 7 feet by 6 feet. Two standard 36-inch propellers drew air through cooling pipes, across the test section, and into an exit cone. The propellers were on the same shaft, which extended outside the tunnel to a 15-horsepower motor. Refrigeration was achieved by passing the air over coils—780 feet of 1.25-inch pipe—that were cooled by a flood system of liquid ammonia. Airbrush nozzles used 80 pounds of compressed air to atomize water into a fine mist and simulate an icing cloud. Designed with the assistance of NACA engineers from the Langley laboratory, the Goodrich tunnel could produce temperatures as low as 0 °F and wind speeds of 85 miles per hour.[19]

The first test in the tunnel took place on 22 August 1930. A Clark Y airfoil, equipped with a prototype two-tube boot, was placed in the test chamber. With ammonia flowing through the coils and the fan turned off, the temperature in the tunnel reached 26 ° after 1.5 hours. The propeller was then started. As the wind speed reached 60 miles per hour, the temperature rose slightly. Water, with a temperature of 43 °, was introduced into the airstream, producing a slush-type ice. The boot successfully removed several coats of ice before the spray nozzles became clogged.[20]

Although the spray system was never able to produce the small droplets found in a natural icing cloud, the tunnel nonetheless proved a useful research tool. The best way to deal with accumulated ice, Goodrich discovered following numerous tests in the tunnel, was to inflate the boots three times a minute. Engineers developed a lightweight air pump that was equipped with a valve that would open automatically and inflate the boots. Other work in the tunnel led to overshoes that would protect tail surfaces, struts and other parts of the airplane that would be vulnerable to icing.[21]

Early in 1931, Goodrich equipped a Lockheed Vega—*Miss Silvertown*—with the protective system that had been developed in the tunnel. Tailored boots were snapped onto the leading edge of the wing, zippered to struts, and laced to tail surfaces. An air compressor installed on the motor automatically supplied air to the inflatable boots. On

---

[18] Geer, "Ice Hazard;" Ben Kastein, "Russell S. Colley, Inventor," *Rubber World,* June 1982, pp. 38–40.

[19] Colley, "Problem #4825: Goodrich Refrigerated Wind Tunnel," 13 September 1930, The Records of BFGoodrich Aerospace, Akron, Ohio. The author is indebted to David Sweet of BFGoodrich's Ice-Protection Systems Division for this memorandum and for other material relating to Goodrich's icing research.

[20] Ibid.

[21] *New York Times,* 29 October 1933.

30 March 1931, Charles Meyers flew *Miss Silvertown* through icing clouds that extended from 2,000 to 8,000 feet over Akron. The system worked perfectly. This flight, the *New York Times* announced the next day, marked "victory" over "one of aviation's most dangerous enemies."[22]

Most airlines quickly adopted the Goodrich boots. TWA equipped its Northrop Alpha 4-As with boots on wings and tails in the winter of 1932–33. In the summer of 1933, United Airlines ordered boots for its Boeing 247s, while TWA put them on its Douglas transports during the winter of 1934–35. The airlines encountered and overcame numerous installation problems. The rubber boots, for example, refused to remain attached to airfoil surfaces at high speeds. Goodrich engineer Russell Colley solved this problem by inventing a hollow threaded rivet called a Riv-Nut that could be installed from outside the wing.[23]

Goodrich, working in cooperation with I. R. Metcalf of the Bureau of Air Commerce and Walter R. Hamilton of TWA, also used the icing tunnel to develop a propeller de-icing system. It consisted of a circular trough of inverted U-section that was bolted to the rear face of the propeller hub and fitted with short tubes that opened at the base of the propeller blade. De-icer fluid, usually a mixture of glycerin and alcohol, was produced at a pressure of 4–9 pounds from a storage tank inside the cockpit and then dripped into the U-section on the propeller hub. Centrifugal force then carried the fluid along the bare aluminum alloy blades. Following a long series of flight tests by TWA pilot D. W. Tomlinson, this "slinger ring" arrangement became standard equipment on the nation's air transports.[24]

By the winter of 1935–36, most airlines had retrofitted their fleets with the modified Goodrich de-icing system. The total weight of the system, at least on TWA's Douglas transports, was 177 pounds, and it cost $65,000. This seemed a small price to pay for the protection against icing. In August 1936, TWA President Jack Frye proclaimed that results during the past winter proved that the new de-icing equipment "has served virtually to eliminate ice formation as a danger to scheduled flight."[25]

Frye's optimistic comment would soon come back to haunt him. On 26 March 1937, the front page of the *New York Times* announced the crash of "a giant Transcontinental Airways skyliner" the previous evening while attempting to land in Pittsburgh. All thir-

---

[22] "Goodrich Airplane De-Icers," *Aero Digest* 18 (May 1931): 66; New York Times, 13 March 1931.

[23] Kastein, "Colley."

[24] S. Paul Johnson, "Ice," *Aviation* 35 (May 1936): 15–19; Jerome Lederer, *Safety in the Operation of Air Transportation* (Norwich University, 1939).

[25] Fred L. Hattoom, "Installation of De-Icer Equipment for Winter Airline Service," *Aero Digest* 29 (November 1936): 38, 86; Jack Frye, "No Ice Today," *U.S. Air Services* 21 (August 1936): 13–14.

teen passengers and three crewmembers had perished when the DC-2 dove nose first into a small gully near Clifton, Pennsylvania, 10 miles from Pittsburgh. It did not take investigators long to establish the cause of the accident. Observers who reached the scene of the crash reported that 1.5 inches of ice remained on the leading edge of the ailerons and on the wing tips of the shattered airplane, which had not burned. A report of the Accident Board of the Bureau of Air Commerce confirmed the initial findings—ice had brought down one of TWA's new transports.[26]

The TWA crash set in motion a chain of events that would bring the NACA back into icing research. The committee had done little in this area since 1931. The small icing tunnel at Langley had rarely been used. The last experiment in the tunnel—a test of an ice-phobic substance—had taken place in August 1935, and there had been no plans for any further use of the facility. The aerodynamicists at Langley did not think highly of the pneumatic boot system, which they believed caused drag. As far as they were concerned, researchers Theodorsen and Clay had demonstrated in 1931 that the application of heat was the answer to the icing problem. The fact that manufacturers had been slow to engineer such a system was not the concern of the NACA.[27]

In the wake of the TWA accident, however, icing became an issue that the NACA could not ignore. On 8 April 1937, Paul E. Richter, vice president in charge of operations for TWA, wrote to Rear Adm. A. B. Cook, chief of the Bureau of Aeronautics, seeking his assistance in persuading the NACA "to proceed at the earliest instant with an investigation of ice formations which must produce solutions to every aspect of the problem." No transport company, Richter emphasized, had "the personnel or facilities to undertake the scientific program necessary to solve this vital problem."[28]

Cook was quick to lend his support to Richter's request. Two weeks later, he informed Dr. Lewis that despite the NACA's earlier work, the problem of icing "has not been completely solved." The NACA, therefore, should "continue these investigations, particularly in reference to formations [of ice] on wings and control surfaces." A successful solution to the icing problem, he emphasized, had both commercial and military value and should be given "the highest priority."[29]

Lewis forwarded Cook's letter to engineer-in-charge Reid at Langley. Reid, in turn, sought the views of Smith J. DeFrance, senior aeronautical engineer at the laboratory.

---

[26] New York Times, 26 and 27 March and 5 May 1937.

[27] William C. Clay to Chief, Aerodynamics Division, 28 August 1935; H. J. E. Reid to NACA, 8 April 1937; Smith J. DeFrance to Engineer-in-Charge, 20 May 1937; all in RA 247, Langley Library.

[28] Richter to Cook, 8 April 1937, RA 247, Langley Library. In his letter to Cook, Richter noted that he had sent a similar request to General A. W. Robbins, commanding officer of the Air Corps' Materiel Division.

[29] Cook to Lewis, 24 April 1937, RA 247, Langley Library.

DeFrance recommended that any icing tests be conducted in flight. A model airfoil could be attached to a research airplane together with a water-spray system so that photographs could be taken of ice formations. Because of the size of the model, the tests could not be done in the existing refrigerated wind tunnel, and he would not recommend "that a larger tunnel be constructed." The water-spray method on an airfoil attached to an airplane, he concluded, would produce better results than tunnel research and would be far more economical. Reid endorsed DeFrance's views, which he passed along to Lewis as the position of the Langley laboratory.[30]

The Langley laboratory, however, would not have the last word on the matter of icing research. Edward P. Warner, influential chairman of the NACA's Aerodynamics Committee, not only wanted the NACA to proceed with an investigation of icing, but also believed that it should be done with the assistance of a new and larger refrigerated wind tunnel. With Warner's strong support, the Aerodynamics Committee recommended that a refrigerated wind tunnel be built. Icing, Warner told his colleagues, was considered by many commercial pilots to be their worst problem. The NACA's Executive Committee endorsed the recommendation of the Aerodynamics Committee, which was then approved by the full Committee on 6 June 1937.[31]

In mid-June, Lewis visited Langley to discuss "the problem" of a large refrigerated wind tunnel. The proposed facility, he told DeFrance and Eastman N. Jacobs, would accomplish two purposes. Not only would it be used for icing experiments, but it would be the model for Jacobs's long-desired low-turbulence variable-density wind tunnel (VDT) as well. Appropriations had been provided in funds for 1939 to build the pressurized two-dimensional, low-turbulence facility that Jacobs needed for his work with laminar-flow airfoils. The Bureau of the Budget, however, required complete plans and specifications for the VDT tunnel by 1 July 1938. Jacobs could use the icing tunnel as his model to secure the necessary information to design the VDT.[32]

Construction of the icing tunnel began early in 1938. W. Kemble Johnson recalls being asked by construction administrator Edward Raymond Sharp to head a major project to build a new wind tunnel at Langley and to modify existing tunnels and other facilities. The money for the project, Sharp said, would come from "post-account" funds

---

[30] DeFrance to Reid, 30 April 1937; Reid to Lewis, 4 May 1937; both in RA 247, Langley Library.

[31] Lewis to Joseph S. Ames, 26 April 1938, RA 247, Langley Library.

[32] Lewis to Ames, 26 June 1937, and 26 April 1938; both in RA 247, Langley Library. James R. Hansen, *Engineer in Charge: A History of the Langley Memorial Laboratory, 1917–1958* (NASA SP-4305, 1987), p. 110, contends that Lewis had always intended the tunnel to be used as a variable-density facility and that labelling it an icing tunnel had been "a necessary political subterfuge." There is no indication of this in the material in RA 247.

and would involve internal resources. The first priority would be the construction of an icing tunnel. Aircraft had been encountering icing problems, Sharp explained to Johnson, and the NACA needed more information about how and when ice formed and what could be done to get rid of it.[33]

Putting up the 100-foot by 200-foot structure to house the icing research tunnel, which would be located behind the Technical Services Building at the Langley laboratory, was no easy task. "We built that from scratch," Johnson remembered; "I mean, we were poor people." He went to nearby Fort Eustis and located steel, trusses, and support columns from a building that had been torn down. The materials were lying in the weeds "with practically trees growing through them." He had welders straighten out the trusses, then cut off the ends of columns and attach them to other columns in order to get the required height for the new building.

For insulation, Johnson secured surplus life preservers from the Navy Yard in Washington, DC. He then hired half the local high school football team to open the pre-servers and fluff up the kapok which would be used to insulate the tunnel. "That was quite a nasty mess," Johnson noted; "[the workers] had to wear face masks and respirators to keep from breathing in dust and fluff from the kapok."

A simple refrigeration system was devised. Johnson purchased carloads of dry ice that would be used to cool an open tank of ethylene glycol. A fan circulated air through the 7-foot by 3-foot test section, while adjustable spray nozzles put water into the airstream.

The cooling system was tried out for the first time on a hot summer's night in 1938. High schoolers were again on hand— this time to chop up the dry ice. A fog of dry ice quickly rose up on the floor of the tunnel to a depth of about 2 feet, with a layer of CO_2 on top. Above that was a 0.5-inch-thick layer of mosquitoes. "It was a very weird thing," Johnson recalled. Nonetheless, the system worked, and the facility had cost only 100,000 depression-era dollars.

In April 1938, as the new refrigerated wind tunnel neared completion, Warner and other senior NACA officials had visited Langley to inspect the work that was being done at the laboratory. As part of their tour, the group was taken to the test chamber of the icing tunnel and treated to a lecture by Jacobs about how the facility would be used as a model for the new two-dimensional tunnel. This came as news to Warner, who promptly expressed his concern to Lewis, "lest the non-turbulent qualities of the icing tunnel obscure the fact that it is an icing tunnel. I appreciate Jacobs's desire to do in that tunnel the work for which he finds it so exceptionally well fitted; air transport, at least, needs a

---

[33] Interview with W. Kemble Johnson by Michael D. Keller, 27 June 1967, copy in the NASA History Office, Washington, DC.

solution to the icing problem just now much more than it needs further increases in efficiency." Warner suggested that transport lines be asked for suggestions about the type of research that would be most urgent for them.[34]

Lewis assured Warner that the work that Jacobs was doing in the tunnel was only for the purpose of obtaining data for the two-dimensional wind tunnel and was not in any way holding up the operation of the icing tunnel, which was set to be ready in July. "I concur with you 100 percent," Lewis told Warner, "in your remarks concerning the icing tunnel and its use." The air transport companies would be contacted, and a schedule of icing research would be drawn up.[35]

Lewis made good on his promise to Warner and solicited suggestions from the airlines through the Air Transport Association. The airlines responded with a lengthy list of possible areas for investigation. American Airlines, for one, envisioned airplane icing tests that would encompass wings, windshields, struts, propellers, control surfaces, airplane skin, engines, cabin windows, exterior lights, and ventilating system inlets and outlets.[36]

As it turned out, the Langley laboratory had its own agenda for tests in the new icing tunnel, and this agenda had little to do with what the airlines wanted to accomplish. Lewis A. Rodert, a junior aeronautical engineer who had joined the NACA in September 1936, took the lead in formulating the laboratory's icing research program. For Rodert, as well as for most of his engineering colleagues at Langley, the answer to the airline industry's icing problems lay in thermal de-icing systems. Existing data indicated that sufficient exhaust heat was available for de-icing. The problem, Rodert noted, was "one of distribution." In order to investigate this aspect of a thermal de-icing system, Rodert wanted to test models in the new icing tunnel. Securing the approval of his superiors, he used three models of 6-foot sections of NACA 23012 airfoils with a 72-inch chord, each with a different duct system. An electric heater and three small electric fans circulated hot air through the ducts. With temperatures from 20° to 28°F and a wind speed of 80 miles per hour, he adjusted the spray nozzles to produce water droplets that varied from 0.002 to 0.05 inches in diameter. He found that ice could be removed or prevented from

---

[34] Warner to Lewis, 23 April 1938, RA 247, Langley Library. Jacobs responded to Warner's criticism. "I must admit," he wrote, "I was discouraged and disheartened over learning Mr. Warner's reaction toward our efforts to advance wind-tunnel technique as indicated by this new equipment. To me, it appears to represent in many ways a successful attempt to keep ahead of foreign countries in our research methods, but evidently to him it inspired a comment no more eloquent than it is an icing tunnel." Jacobs to Chief, Aerodynamics Division, n.d. [May 1938], RA 247, Langley Library.

[35] Lewis to Warner, 28 April 1938, RA 247, Langley Library.

[36] Lewis to Edgar S. Gorrell, 7 May 1938; William Littlewood, vice president, engineering, American Airlines, to Fowler W. Barker, 20 May 1938; both in RA 247, Langley Library.

forming by heating the skin of the leading 10 percent of the airfoil to a temperature that was approximately 200 degrees above tunnel air temperature. The required gas temperature in the duct to produce this skin temperature varied from 360° to 834°F.[37]

Although Rodert also used the icing tunnel to conduct experiments with electrically heated windscreen panels, he saw no future for tunnel-based research. He believed icing tunnels could not create conditions that resembled natural icing and far more could be accomplished with flight research. His colleagues at Langley agreed. When the Glenn L. Martin Company asked for the NACA's assistance to run icing tests on a cowling, engineer-in-charge Reid recommended that the request be turned down. The icing tunnel, he wrote to NACA headquarters in the fall of 1939, was not being used for icing research "because of the fact that we are obtaining excellent results in our flight tests, which are being pushed at this time by all the personnel available." Also, Reid continued, the icing tunnel "is in constant use in the further study of low-drag airfoils." The Langley laboratory had obviously won the battle over the use of the refrigerated wind tunnel.[38]

As Reid had noted, Rodert had moved on to flight tests of thermal systems. Initially, this involved mounting a model of a NACA 0012 airfoil, with a span of 4 feet and a chord of 3 feet, between the wings of an XBM Navy biplane. He wanted to determine the amount of heat that could be extracted from the exhaust gas of the airplane's engine. An exhaust tube was placed inside the model along the interior of the leading edge to carry the hot gas. Flying in temperatures between 17° and 25°F, he turned on a spray nozzle that was mounted in front of the airfoil to test both ice prevention and ice removal. The system, he found, could melt 0.5–1 inch of ice in 10 to 30 seconds.

"In view of the favorable results of the NACA investigations on the application of heat in de-icing," Rodert concluded, "and also in consideration of the reports that have been received describing the successful application of exhaust-heat de-icing on numerous four-engine transport airplanes in Germany, it is believed that full-scale application of this method should be undertaken at an early date in the United States."[39]

Upon his recommendation, NACA Headquarters approved the purchase of a Lockheed 12A for icing studies. Working with chief engineer Hall L. Hibbard of

---

[37] Rodert, "A Preliminary Study of the Prevention of Ice on Aircraft by Use of Engine-Exhaust Heat," NACA TN 712 (June 1939).

[38] Reid to NACA, 23 February and 11 September 1939, RA 247, Langley Library.

[39] Rodert and Alun R. Jones, "A Flight Investigation of Exhaust-Heat De-Icing," NACA TN 783 (1940). Rodert obtained information on German exhaust-heating deicing in 1939. See BuAer to NACA, 29 June 1939, RA 247, Langley Library. He later had an opportunity to examine and admire the anti-icing systems of a Junkers JU-88. See Rodert and Richard Jackson, "A Description of the Ju 88 Airplane Anti-Icing Equipment," NACA Restricted Bulletin, September 1942.

Lockheed, Rodert and Alun Jones modified the aircraft with a "hot wing," using valves in the engine exhaust stack to divert hot gas into a 4-inch-diameter tube that ran close to the leading edge. By the time the aircraft was ready in January 1941, the NACA had transferred Rodert and his icing research program to the new Ames Aeronautical Laboratory in Moffett Field, California. During the war years, Rodert and his team worked to perfect a thermal de-icing system, first with the Lockheed and later with a B-24 and a B-17. In 1943, he converted a Curtiss C-46 into a flying laboratory and continued the flight-test program at the Army Air Forces's Ice Research Base at Minneapolis. Although Rodert opposed the use of chemically heated air for his thermal de-icing system, aircraft manufacturers distrusted the use of exhaust gas, which they feared would corrode aluminum and was potentially dangerous. At the end of the war, many manufacturers turned to chemically heated air for their de-icing systems. The Douglas DC-6, for example, used gasoline-burning heaters that were built into the engine nacelles. Rodert's work, nonetheless, was hailed as pioneering and would be recognized by the award of the 1946 Collier Trophy.[40]

Rodert and the aerodynamicists at the Langley laboratory never had any confidence in the utility of a refrigerated wind tunnel as a research tool, but others in the NACA were not as pessimistic. Indeed, while Rodert was conducting his flight experiments at Ames, work was beginning on another icing tunnel at the new NACA Aircraft Engine Research Laboratory in Cleveland.

---

[40] See Glenn E. Bugos, "Lew Rodert, Epistemological Liaison, and Thermal De-Icing at Ames," in Pamela E. Mack, *From Engineering Science to Big Science: The NACA and NASA Collier Trophy Research Project Winners* (Washington, DC: NASA SP-4219, 1998), pp. 29–58.

Chapter 2
The IRT Takes Shape

Existing documentation does not include a record of the discussions which led to the NACA's decision to build a new refrigerated icing tunnel after the Langley facility was converted to other use, but the sequence of events is clear. In December 1938, in the face of an increasingly tense international situation, the NACA responded with plans to build two new research laboratories. One would be located on the West Coast and would emphasize flight research. The other would be built in the central part of the country and would focus on engine research. Congress approved funding for the two facilities in June 1940, and construction began for the new NACA Aircraft Engine Research Laboratory (AERL) in Cleveland in January 1941.[1]

The centerpiece for the AERL would be an Altitude Wind Tunnel (AWT), which would allow engines to be tested at high altitudes and high speeds. The AWT would require a revolutionary refrigeration system to simulate the low temperatures found at high altitudes. It was the presence of this system, with its excess capacity, that permitted the possibility of building a tunnel next to the AWT that could be used for icing research.

The first document that discusses an icing research tunnel is dated 16 February 1942. It is clear that a decision in favor of constructing what became the Icing Research Tunnel (IRT) had been made by that point in time, but that opposition remained. A conference was held on that date with three Langley engineers: John W. Crowley, Jr., chief of research; Carleton Kemper, head of engine research; and Robert T. Jones, a distinguished aerodynamicist. Crowley argued—as Langley aerodynamicists had been arguing for a decade—that the icing problem had become one of engineering, not research. The icing tunnel at Cleveland should therefore be designed with aerodynamic requirements in mind. Kemper agreed. Although there was a need for icing tests of scoops, radiators, pitot tubes, and other components, the tunnel

---

[1] On the AERL, see Virginia P. Dawson, *Engines and Innovation: Lewis Laboratory and American Propulsion Technology* (Washington, DC: NASA SP-4306), 1991.

Figure 2–1. General George Brett, with shovel, and Dr. George W. Lewis, with pick, break ground for the AERL, 23 January 1941. (NACA C-88-3574)

should not be limited by design to icing tests. Jones believed that some types of icing research were still needed because of the limitations of flight testing. The three men finally agreed "that the most desirable course of action would be to take the present 7 x 10 tunnel design of Moffett Field and adapt same with a minimum of change for the Cleveland Ice Tunnel."[2]

Eleven days later, NACA Langley wrote the design specifications for a large refrigerated wind tunnel. The tunnel, it began, should simulate flight conditions as closely as possible. The test section, which should permit the testing of at least medium-size wings, should be 10 feet wide and 7 feet high. The tunnel should be capable of producing wind speeds of at least 300 miles per hour and temperatures as low as –20°F. In order to save time in designing the tunnel and preparing the necessary working drawings, the Cleveland facility could use the plans for the 7 x 10-foot tunnel at Ames, with

---

[2] Charles N. Zelenko, "Cleveland Ice Tunnel: Conference with Crowley, Kemper and Jones," 16 February 1942, History Office, Glenn Research Center (GRC), Cleveland, OH.

Figure 2–2. Construction begins for the AERL, 8 May 1942. (NACA C-8289)

only minor modifications. Also, the tunnel could be used for aerodynamic research in the event that enemy air attack damaged the NACA's coastal laboratories.

Turning to specifics, the plans called for a closed, continuous airtight passage through which air would be circulated by a six-bladed propeller, driven by an electric motor. The motor would be placed in a faired nacelle and located in the airstream. The facility would cover an area of approximately 230 feet by 85 feet, with tunnel passage varying from 30 feet by 33 feet at maximum cross-section to 7 feet by 10 feet at the test section. The test chamber, 45 feet long, would include a hinged access door at the top and observation windows at the sides. Refrigeration would be supplied by a heat exchanger that would consist of banks of four or five rows of round pipes, evenly spaced, with a gap of 2 feet between each bank. The plans were vague about the spray system in the tunnel, merely noting that a "means" would be provided for injecting water into the air stream. The tunnel should cost approximately $690,000, a figure that was based on the cost of the Ames tunnel, plus adjustments.[3]

---

[3] NACA, Langley, "Design Specifications for Cleveland Ice Tunnel," 27 February 1942, History Office, GRC.

Using Langley's specifications as a basis for their plans, the design and construction team at Cleveland began work on the tunnel in the spring of 1942. Four individuals played key roles in the creation of the new facility. Edward R. Sharp, as construction administrator (later, manager) of AERL, was in overall charge of the project. His chief lieutenants were Ernest G. Whitney, head of the design group at AERL, and mechanical engineer Alfred W. Young. Charles N. Zelenko, an assistant aeronautical engineer, had responsibility for translating plans into reality.

From the beginning, the design team at AERL had questions about Langley's original specifications. Economic factors argued strongly for a somewhat smaller facility than the 7-foot by 10-foot tunnel. Information from Goodrich had pointed out that the starting point for the design of any icing tunnel should be the selection of the maximum water density to be sprayed across the throat. This spray places the heaviest load on the refrigeration system, causes ice deposits on turning vanes, creates ice removal problems, and causes possible icing of the heat exchanger. When simulating icing conditions at 300 miles per hour, the maximum refrigeration load for a 7 x 10-foot tunnel would be approximately 2,000 tons. Achieving this would require a total of 8,645 horsepower per hour. With Cleveland's high power costs, this would translate into $25.80 per hour.

If the tunnel could be scaled down to 6 feet by 6 feet, the savings would be considerable. Goodrich indicated that the smaller tunnel would be suitable for icing experiments. The construction costs would go down from $650,000 to $458,000. The smaller tunnel would require only 4,490 horsepower to produce simulated icing conditions, lowering operating costs to $13.40 per hour.[4]

In July 1942, Young and Whitney approved an amended set of design specifications for the tunnel. The overall shell would be 200 feet by 75 feet, containing a test section that would be 6 feet by 9 feet by 25 feet long, which would permit the testing of at least medium-size wings. Its "basic throat" would accommodate wind speeds of 300 miles per hour, with a smaller auxiliary throat that would allow speeds of 400 miles per hour. "The tunnel," noted the specifications, "shall be a low turbulence tunnel suitable for aerodynamic tests."

As originally planned, the tunnel would be a closed continuous passage facility through which air would be circulated by means of a propeller driven by an electric motor that would be placed in a faired nacelle. A temperature of -40°F would be achieved by using the refrigeration equipment of the Altitude Wind Tunnel. As in earlier specifications, the nature of the spray system was left vague. The construction cost for the 6 x 9-foot tunnel was estimated at $559,138.[5]

---

[4] Zelenko to Whitney, "Design Consideration for the Refrigeration Tunnel," 24 March 1942, History Office, GRC.

[5] Zelenko, "Design Specifications for Ice Tunnel," 9 July 1942, History Office, GRC.

Figure 2–3. Shell framing for the Icing Research Tunnel, with Altitude Wing Tunnel in background, 21 July 1943. (NACA C-IRT-4)

Construction of the basic tunnel posed no particular problems for the NACA, as the committee had extensive experience in building wind tunnels. The Pittsburgh-Des Moines Steel Company had been selected as the general contractor for both the Altitude Wind Tunnel and Icing Research Tunnel, and the company's only challenge related to wartime priorities for materials. The real challenge came with the design and construction of the refrigeration unit that would be used for both tunnels.[6]

When Willis Haviland Carrier, known as the "father of air conditioning," saw the original plans for refrigerating the NACA facility, his reaction was "impossible." The NACA wanted 10 million cubic feet of air per minute to be cooled to -67°F for the Altitude Wind Tunnel. Nothing of this magnitude had ever been done before. Furthermore, Carrier believed that NACA engineers would fail in their

<hr>

[6] Pittsburgh-Des Moines Steel Company, "Invitation for Bids for Ice Tunnel Heat Exchanger and Equipment," 21 September 1942, History Office, GRC.

attempts to accomplish this cooling objective by using an experimental coil with streamlined tubes.[7]

Carrier recalls a luncheon meeting in Washington, DC, with Drs. Vannevar Bush, Jerome Hunsaker (chairman of the NACA), and George Lewis. At the time, NACA engineers were testing the streamlined coils at Langley. Lewis asked Carrier if the experimental coils had any value. Carrier's reply was blunt. "I told Dr. Lewis," he remembered, "that the boys conducting the tests did not know what it was all about, and that too much money and, of more importance, too much time had been wasted already." It was Carrier's outburst, together with disappointing results from early tests with the streamlined coils, that likely led to a meeting with Carrier, representatives of his company, and high officials from Langley on 6 November 1941. At the beginning of the meeting, Carrier announced that his company had decided not to bid on the contract for the refrigeration plant at Cleveland. As an executive of the Carrier Corporation, L. L. Lewis later explained, the company was "loaded with work in familiar fields," whereas the NACA project would involve "great risk and great effort." It was suspected that Willis Carrier, having been ignored by the NACA at the outset of the project, now wanted to be courted.[8]

Russell G. Robinson, a senior NACA official, was prepared to be the suitor. "I explained the urgency of the project in the interests of national defense," Robinson said, "and pointed out that the highest priorities could be obtained wherever advantageous." The end of the day's discussion had persuaded Carrier. His company would bid on the project. Robinson was pleased. "I was impressed," he wrote, "by the confidence with which Carrier [Corporation] approached this problem; they seem entirely capable of carrying out a project such as ours."[9]

After its bid was approved in March 1942, the Carrier Corporation created a special department, headed by Maurice J. Wilson, to undertake the challenging task. There were several major problems to be overcome to create the cooling system for Cleveland. "Calculations indicated," Willis Carrier pointed out, "that we would need a direct expansion coil with a face area of approximately 8,000 square feet." But the wind tunnel, 51 feet in diameter, had only 2,000 square feet of cross-sectional area. Carrier solved this problem by folding the coils "like a collapsed accordion until they fitted into the tunnel."[10]

---

[7] Margaret Ingels, *Willis Haviland Carrier: Father of Air Conditioning* (Country Life Press, 1952), pp. 96–101.

[8] Ibid.

[9] Russell G. Robinson, "Conference with Representatives of Carrier Corporation," 6 November 1941, History Office, GRC.

[10] Ingels, *Carrier.*

Figure 2–4. Construction of balance frame supporting piers for the IRT, with ventilating tower framing in left background, 5 August 1943. (NACA C-IRT-9)

A system to supply refrigerant to the coils also challenged Carrier's engineers. Freon-12 had been specified as the refrigerant needed to produce a temperature of -67°F. But Freon-12 had never been used in such a large system. Engineers redesigned their centrifugal compressors to handle a maximum of 7,150 tons of Freon-12.[11]

"Much was not standard," Willis Carrier observed, "nor could it be, for such an unusual installation." After numerous shakedown difficulties and false starts, the system was ready on 24 April 1944 for a formal run-in test. Carrier was present to watch the innovative system come to life. It worked as designed. Carrier later was to hail this system as his company's greatest engineering feat.[12]

In the summer of 1943, as work on the IRT and the refrigerating system continued, AERL created a new division to conduct icing research and hired Willson H. Hunter to take charge of it. Hunter brought extensive icing experience to the task. After

---

[11] L. L. Lewis, "Carrier in World War II Wind Tunnel Air Conditioning," 26 September 1956, Records of the Carrier Corporation, 1875–1964, Box 16, Cornell University Library; Ingels, *Carrier*.

[12] Ingels, Carrier; L. L. Lewis to E. T. Murphy, "Wind Tunnel Activity," 21 April 1944, and W. H. Carrier to Maurice J. Wilson, 3 March 1945, Records of the Carrier Corporation, Box 16.

receiving his bachelor's degree in mechanical engineering from Yale University in 1930, Hunter had joined the Goodyear Zeppelin Corporation in Akron as a power plant design engineer for the Navy's rigid airships *Akron* and *Macon*. Four years later, he went to work for the BFGoodrich Company as an aircraft wheel and brake development engineer. In 1938, he became aeronautical research supervisor, and later manager, of de-icer research and development.[13]

Hunter's first task at AERL was to design the spray system for the IRT. While using nozzles similar to the ones employed in the Goodrich icing tunnel, he came up with a novel way to arrange them. His aim was to concentrate the spray around the model that was to be tested and thereby reduce icing in the tunnel to a minimum. The Goodrich tunnel had nozzles mounted on horizontal spray bars. Working with Halbert E. Whitaker, Hunter had the Cleveland nozzles mounted on a rotating air-foil that was located 30 feet in front of the test section. There were in fact two sets of nozzles. One set consisted of the end of a 0.25-inch-diameter stainless steel tube that had been flattened to a gap of 0.01 inches and would produce a fine spray. Behind the water nozzles were air nozzles, which were supposed to regulate the size of the water droplets and the dispersal of the spray. At the time, Hunter's arrangement was hailed as "ingenious."[14]

Calibration tests for the IRT began on 22 March 1944. To accomplish this task, a metal grid was placed in the center of the test section, and the tunnel was brought to a given temperature and airspeed (about 30° and 200 miles per hour). The tunnel oper-ator then would set the water and air pressure of every spray nozzle to produce a droplet size and liquid water content that would result in a uniform accretion of ice on all por-tions of the grid. If successful, tunnel operators would know that these nozzle settings under these conditions would produce the uniform icing cloud that would be found in nature. Different temperature and airspeed combinations would require different air- and water-pressure settings; and nozzles, despite their similar design, had individual peculiarities. Calibration was a tedious but essential process.

From the beginning of the calibration runs, it was clear that Hunter's spray system had problems. Distribution of the icing cloud was uneven. The spray tended to concen-trate in the middle of the test section, with little if any water around the periphery. Whitaker tried several commercially available spray nozzles to correct the situation. While the new nozzles tended to produce a more uniform cloud, the water droplet size was far too large. Also, the nozzles kept plugging up. It turned out that the system's

---

[13] Undated obituary [1983] for Willson H. Hunter, NASA History Office.

[14] J. K. Hardy, "The N.A.C.A. Icing Tunnel at the A.E.R.L. - Cleveland," 17 December 1943, NASA History Office.

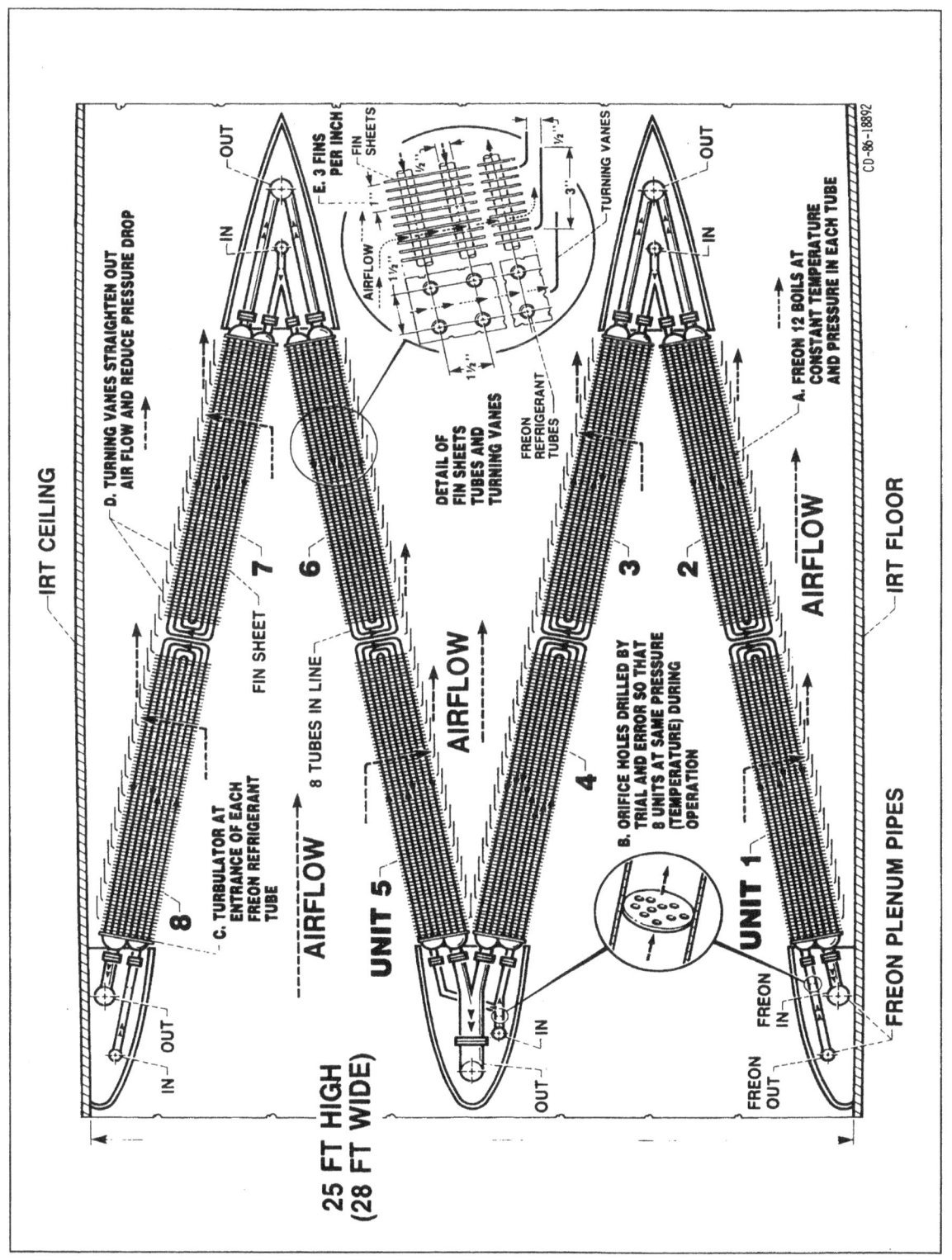

Figure 2–5. Side view of the refrigeration heat exchanger in the IRT. (NACA CD-86-18892)

plumbing had used copper with galvanized steel; this caused a battery reaction that contaminated the spray water and plugged up the nozzles. Engineer Harold Christenson had to tear out the original plumbing due to corrosion and substitute copper plumbing with a stainless steel two-stage pump to keep the water clean.[15]

Adding to Hunter's woes was a major explosion in the tunnel. On the morning of 29 December 1944, W. F. Morse of General Electric was testing the Phantom 4,160-hp motor that powered the tunnel's twelve-bladed fan. During the shutdown procedure, the motor exploded. Apparently, the explosion was caused when a spark from the short circuit area ignited fumes from the insulation of an overheated rotor. AERL's fire department quickly responded. No IRT personnel were injured, but several firemen suffered minor burns and bruises. The motor suffered extensive damage and had to be scrapped. It took two months to repair the damage to the IRT.[16]

Even before the explosion, the first experiments had taken place in the new tunnel. On 31 August 1944, work had begun on bare-blade propeller icing. A second project, lasting from 27 November to 1 December and using 29 hours and 24 minutes of tunnel time, involved antenna icing. After the tunnel reopened following the motor explosion, aerodynamic and icing tests were conducted on C-46 engine air scoops from 27 February to 21 April 1945. Another project, lasting from 21 June to 8 October, involved tests of electrically heated propellers. In all, the IRT was used for a total of 405 hours from 22 March 1944 through 30 June 1945. Research projects accounted for 264 hours and 32 minutes, while 140 hours and 28 minutes were devoted to calibration and demonstrations.[17]

Among these early tests in the IRT, one of the most significant involved the development of a protected air scoop for C-46s. The Army Air Forces had been experiencing high losses of C-46s that were flying the treacherous "Hump" air route between India and China, especially during the monsoon season. At the high altitudes that were frequently required to cross the Himalayas, the pilots had little engine heat to spare to protect the induction system of their engines against icing. Even the slightest loss of power meant a loss of altitude amongst the towering mountains, creating a hazardous and often fatal situation.

Researchers Uwe von Glahn and Clark E. Renner tested C-46 air scoops in the IRT between 27 February and 21 April 1945. They installed the upper half of a C-46 engine cowling in the test section of the tunnel and conducted experiments of a standard scoop

---

[15] Halbert E. Whitaker to William Olsen, 6 March 1986, copy provided to the author by Mr. Whitaker; interview with Harold Christenson by William M. Leary, 22 September 2000.

[16] Edward R. Sharp, "Report of Accident in Icing Tunnel, 29 December 1944," 3 February 1945, History Office, GRC.

[17] Carleton Kemper to NACA Headquarters, 6 November 1945, History Office, GRC.

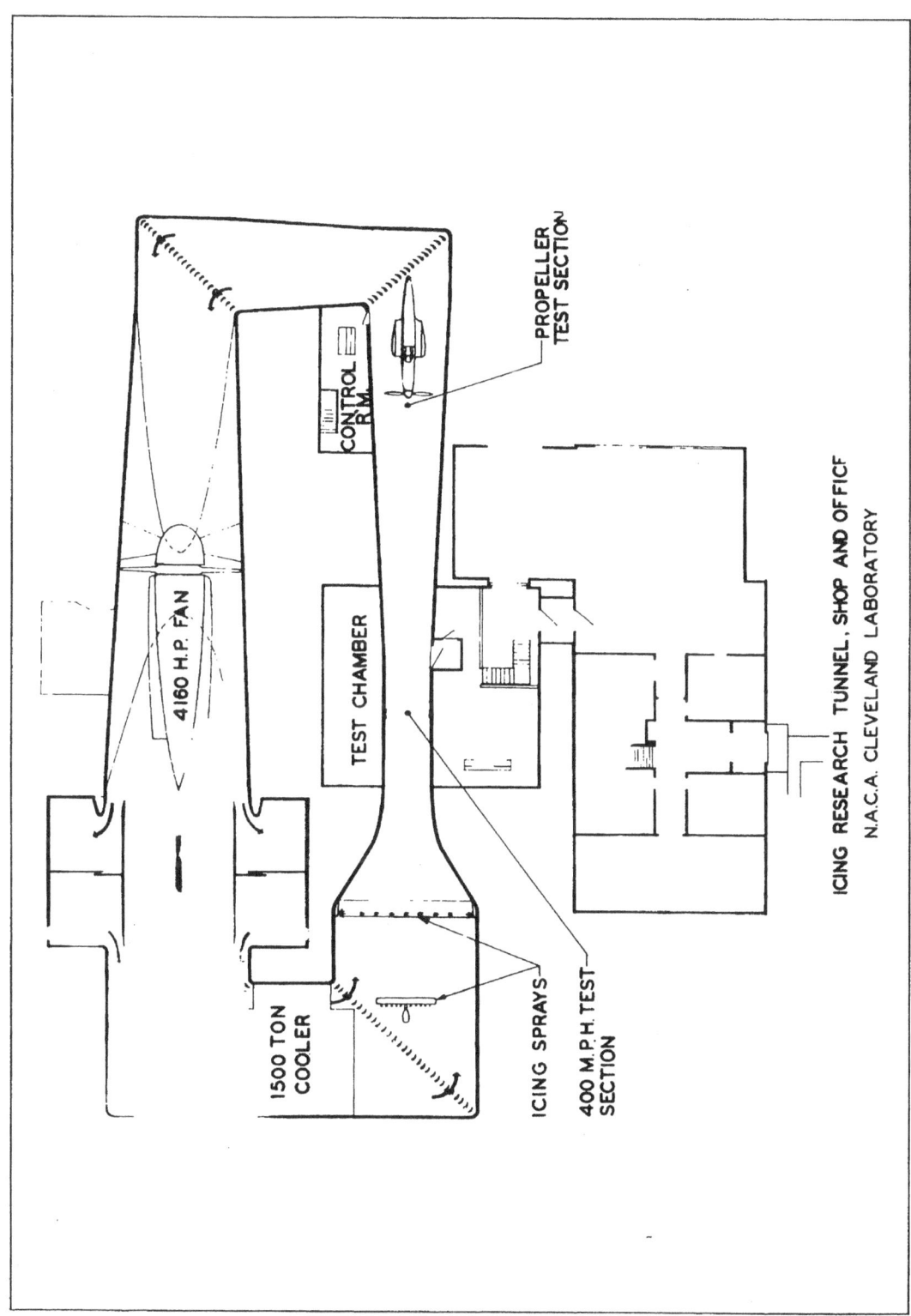

Figure 2–6. Schematic of IRT in 1944–45, showing two of the early spray systems. The upstream spray bar is essentially a propeller with spray nozzles on it. The spray bar at the entrance to the bell mouth has the spray nozzles around the tunnel walls. Neither system worked satisfactorily. (NACA C-11274)

and three modified scoops at flight angles of 0°, 4°, and 8°. There were two sets of tests, one for water ingestion and the other for icing.

During their 120 hours of testing, von Glahn and Renner discovered that the problem with the C-46s was more one of water ingestion than ice. Scoops designed by Willson Hunter and Leo B. Kimball effectively prevented the entry of rain by means of inertia separation. Although these results came too late to be useful for the India-China air route, they provided the basis for the redesign of C-46s, and later Convair 240s, that would be used after the war by the military services and commercial operators.

The icing tests of the C-46 scoops were less successful. As the researchers noted in their report, "true cloud conditions were not simulated in the tunnel." In most natural impact-icing conditions, they pointed out, the diameter of cloud droplets varied between 10 and 30 microns, while the water content ranged between 0.5 and 1.5 grams per cubic meter. The size of the drops in freezing mist and drizzle varied between 50 and 400 microns, while freezing raindrops could attain a size of 2,000 microns. The spray system in the IRT could produce droplets whose sizes were intermediate between an icing cloud and freezing rain. The NACA obviously had more work to do on the IRT's spray system.[18]

Despite the problems with replicating a natural icing cloud, the IRT was kept busy with a variety of projects, most of which were generated by the military services. In 1945, for example, the Air Material Command of the Army Air Forces asked the NACA to test a promising thermal-pneumatic wing de-icing system. The standard pneumatic boot system had several drawbacks. A certain amount of ice had to be tolerated in order to permit cyclic removal. This inter-cycle ice created drag and was difficult to remove. The Air Material Command hoped that a thermal-pneumatic system would eliminate the inter-cycle ice without requiring a large power supply.

The system tested in the IRT featured an electrically heated strip, 3 inches wide and 33.5 inches long, that was cemented symmetrically about the leading chord line of the airfoil. Thirty-two inflatable tubes extended span-wise, sixteen on either side of the airfoil, adjacent to the electrically heated area. The tubes, made from neoprene-covered stretchable nylon fabric, were 0.75 inches wide and extended rearward to approximately 25 percent chord. They required an air pressure of 25 to 30 pounds for inflation.

Researchers William H. Gowan, Jr., and Donald R. Mulholland installed the system on a NACA 0018 airfoil, having a chord of 42 inches and a span of 36 inches. They mounted the airfoil from the ceiling of the test section, approximately 11 feet downstream of a spray strut that atomized water to produce an icing cloud. The

---

[18] von Glahn and Renner, "Development of a Protected Air Scoop for the Reduction of Induction-System Icing," NACA TN 1134 (September 1946); interview with von Glahn by William M. Leary, 17 June 2000.

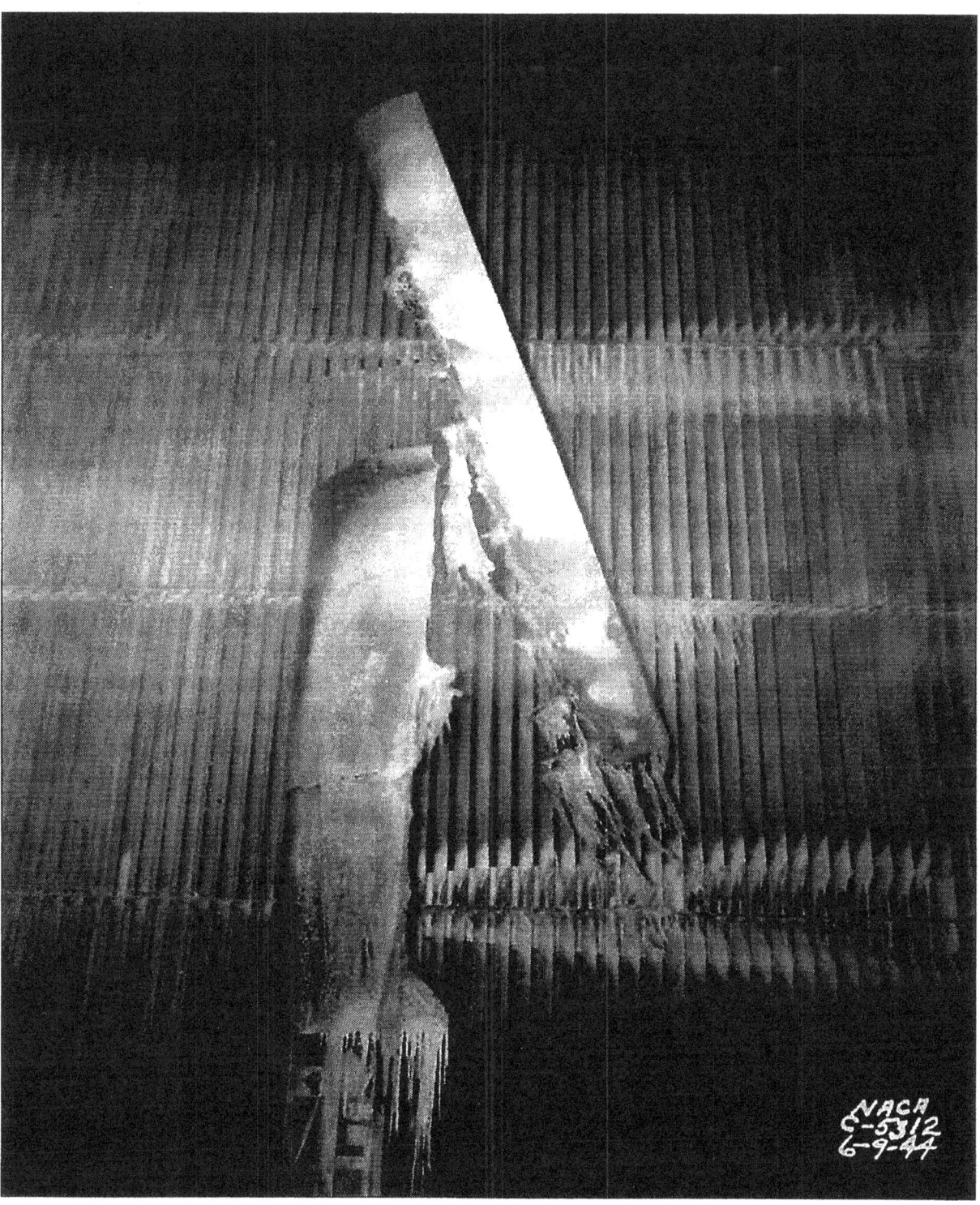

Figure 2–7. Willson Hunter's shirling spray arm, 9 July 1944. (NACA C-5312)

researchers first conducted aerodynamic tests on the model, with and without the de-icer installed, to obtain drag measurements. For the icing runs, they used angles of attack of 0° and 4°, with airspeeds of 150 to 300 miles per hour and air temperatures of 0°, 10°, and 20°F. The icing cloud contained an effective droplet size of approximately 35 microns and a liquid water content of 2 grams per cubic meter. Ice covered the entire area of the de-icer system, Gowan and Mulholland noted, "and is considered an extreme maximum icing condition, which probably would be experienced in the atmosphere only intermittently." This condition, thus, could be considered "a severe test" of the de-icer.

The aerodynamic tests showed that when the de-icer was installed with no tubes inflated, drag increased 140 percent over a bare airfoil. This increased to a maximum of 620 percent with tube inflation. The thermal-pneumatic system, however, when operated on a 2-minute inflation cycle and with continuous electric heating of the leading edge, prevented excessive ice formation. Small scattered residual ice remained after inflation, but was removed intermittently during later cycles. The system showed a good deal of promise and merited further development.[19]

Another major project in the tunnel involved experiments on two systems that were designed to prevent ice from forming on propellers. The slinger rings currently in use, which delivered de-icing fluid to the propellers by means of centrifugal force, worked under only limited conditions. The application of heat held out better possibilities.

Researchers Vernon E. Gray, Donald R. Mulholland, and Porter J. Perkins tested three types of gas-heated hollow-steel propeller blades that had been mounted on a cut-off P-39 Bell Airacoba fuselage in the first diffuser section of the tunnel. To simulate icing, they used a ring of water spray with an average droplet size of 55 microns and an average liquid water content of 0.5 grams per cubic meter. Temperatures in the tunnel ranged from -9° to 23°F. They found that a minimum heat input of 40,000 Btu per hour per blade would afford adequate ice protection.[20]

Researcher James P. Lewis used the same test apparatus to examine two electrothermal systems for propeller ice protection. One method used external rubber-clad

---

[19] Gowan and Mulholland, "Effectiveness of Thermal-Pneumatic Airfoil Ice-Protection System," NACA RM E50K10a (1951). The authors noted that the research, which was done in 1945, was being published in 1951 "because of the inquiries that have been received regarding this type system."

[20] NACA TN 1586 (May 1948), TN 1587 (May 1948), and TN 1588 (May 1948); Vernon H. Gray, "Propeller Ice Protection by Means of Hot Gases in Hollow Blades," NACA Conference on Aircraft Ice Prevention, 26–27 June 1947, pp. 119–27.

blade heaters with small chromel heating ribbons, while the other featured internal electro-thermal heat. Tests in the IRT showed that both systems worked.[21]

On 23 January 1947, the focus of icing research shifted dramatically. Investigation of propellers came to an end. "Our experimental work on the simple air-heated blades," Hunter noted, "has been completed to the point where reports can now be written on each of the three-blade configurations." Effective immediately, he informed his staff, the schedule of the IRT would be changed to expedite research on jet engine inlet ice protection.[22]

Hunter was responding to the demands of the new jet age. As Arthur A. Brown of Pratt & Whitney stated, icing of turbine air induction systems posed "a very serious problem" for the industry. Engine manufacturers needed basic research on the inertial separation of water and ice particles as a means to prevent icing. Some local surface heating may be necessary, but this would impose a large handicap on overall performance. Experiments in Pratt & Whitney's laminar flow wind tunnel had proved inconclusive because models had been tested at low airspeeds with only rough control of droplet size. The results of an investigation by the NACA would be of "tremendous value in our efforts to design and build non-icing turbine power plant installations."[23]

The NACA launched a three-phase investigation into the problem. It began with flight research. A Westinghouse 24C-2 turbojet engine with a ten-stage axial flow compressor was mounted below the wing of a B-24 bomber for tests in natural icing conditions. In March 1948, the B-24 spent one hour in an icing environment with the turbojet engine operating at a speed of 9,000 rpm. The liquid water of the icing cloud was highest during the first 15 minutes, reaching a peak of 0.38 grams per cubic meter. After 45 minutes in the cloud, tail pipe temperature had increased from 761° to 1,065°, while thrust had decreased from 1,234 to 910 pounds. The engine did not have to be shut down, researcher Loren W. Acker noted, but a reduction in engine speed would have been mandatory if it operated at a takeoff power of 12,000 rpm.

A second flight took place in April, this time with the turbojet engine operating at a normal cruising speed of 11,000 rpm. Again, the B-24 spent an hour in a natural icing environment. Upon entering the icing cloud, there was a sudden drop in engine thrust as ice collected on the cowl lip, disrupting airflow and destroying pressure recovery at the com-

---

[21] NACA TN 1520 (February 1948) and TN 1691 (August 1948); Lewis, "Electro-Thermal Methods of Propeller Ice Protection. I - Cyclical De-Icing by External and Internal Blade Heaters," NACA Conference on Aircraft Ice Prevention, pp. 128–36.

[22] Hunter, "Revised Schedule for Icing Research Tunnel to Expedite Jet Engine Inlet Icing Research," 23 January 1947, History Office, GRC.

[23] Brown, "Report of Pratt & Whitney Aircraft Subcommittee on Icing Problems," 29 April 1948, NASA History Office.

pressor inlet. After 18 minutes, however, a low liquid water content of 0.077 grams per cubic meter and a high free air temperature of 25°F caused the inlet ice to melt and allowed the engine to return to normal operation. Seventeen minutes later, the liquid water content rose rapidly to 0.49 grams. Fuel flow had to be increased to maintain engine speed, resulting in an increase in tail pipe temperature and a rapid decrease in thrust from 1,950 to 1,700 pounds. After 45 minutes in the cloud, the engine was accelerated to the takeoff power of 12,000 rpm and held there for 2 minutes. At this point, the tail pipe temperature stabilized. "No general conclusions may be drawn from these data," Acker warned, "but no serious reduction in engine performance would be experienced in a icing condition similar to the one discussed."[24]

In the next phase of the research program, William A. Fleming mounted the Westinghouse engine in a wing nacelle in the test section of the Altitude Wind Tunnel. The engine was equipped with an experimental ice protection system that bled hot gas from the turbine inlet and injected it into the air stream ahead of the compressor inlet, heating the air to above freezing. Tests were conducted at simulated altitudes of 5,000 and 20,000 feet, with air temperatures that ranged from 0° to 35°F.

In an effort to simulate icing, Fleming placed a tower with five spray nozzles a distance of 7 feet in front of the duct inlet. The nozzles, which had been designed by Hunter, injected water into a supersonic airstream through holes that were 0.010-inch in diameter. Hunter acknowledged that the spray system represented only "an intermediate makeshift design," but he hoped to obtain droplets that were 10 to 15 microns in diameter. Unfortunately, his system failed to achieve this objective. While the water spray did not replicate natural icing, researcher Fleming reported, the system was adequate for his experiments.

Fleming found that ice formed so rapidly at the compressor inlet under severe icing conditions that the engine would flame out within 1 to 2 minutes. A hot-air bleedback system could be used to prevent ice formation, but under severe conditions it would require that some 4 percent of the gas be bled to the engine inlet. This would reduce thrust by 18.8 percent (at 12,000 rpm) and increase fuel consumption by 16.5 percent. Obviously, additional research would be required to develop a less costly de-icing system.[25]

The third phase of the investigation was conducted in the IRT by researchers von Glahn, Edmund E. Callaghan, and Vernon H. Gray. The inlet guide vanes, previous studies had revealed, posed the greatest danger to the operation in icing conditions of an axial-flow tur- bojet engine. The researchers set out to test three systems that would protect the vulnerable vanes: surface heating, hot-gas bleedback, and inertia-separation inlets. Local heating could be

---

[24] Acker, "Natural Icing of an Axial-Flow Turbojet Engine in Flight for a Single Icing Condition," NACA RM E8F01a (1948).

[25] Fleming, "Hot-Gas Bleedback for Jet-Engine Ice Protection," NACA Conference on Aircraft Ice Prevention, pp. 86–92.

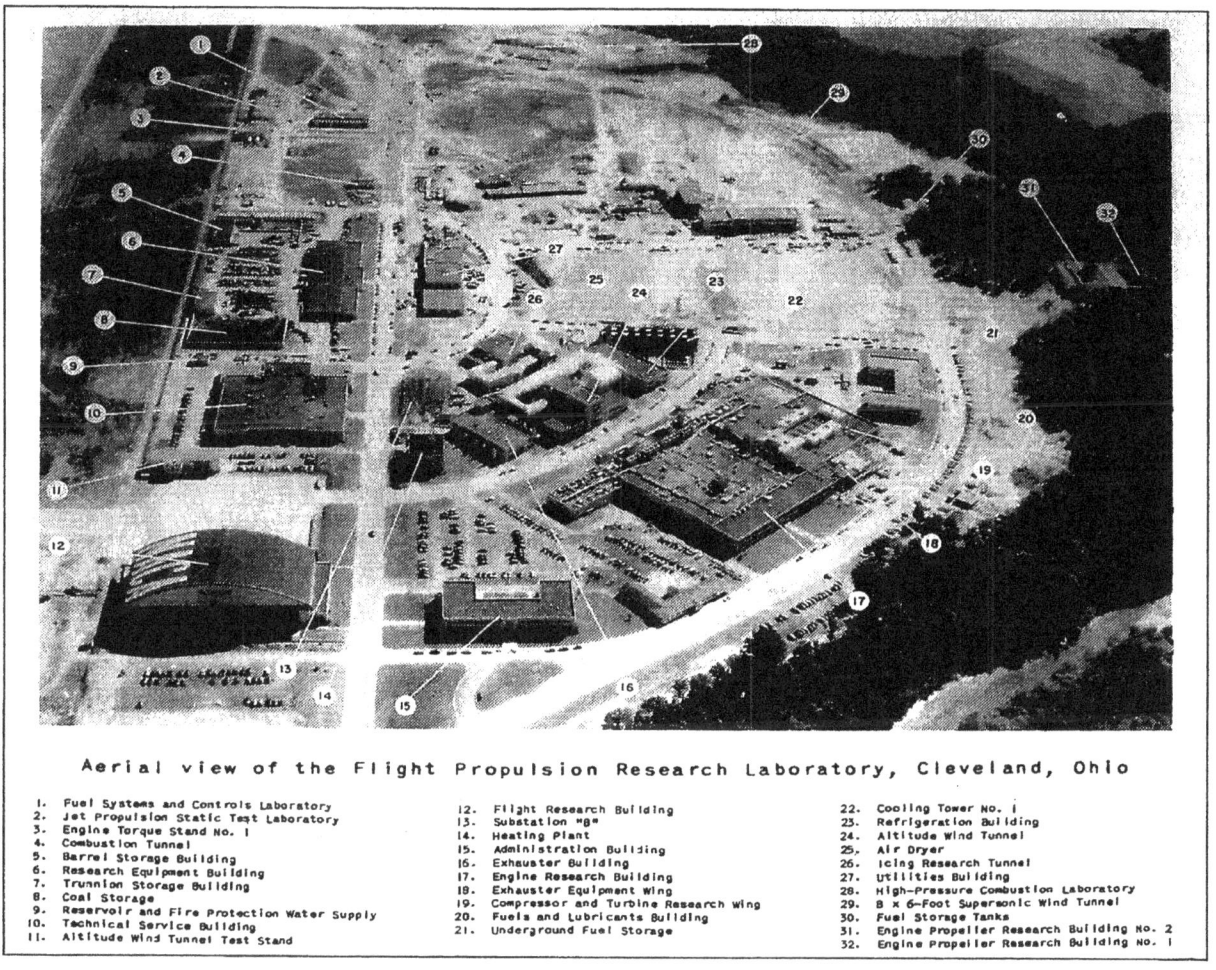

Aerial view of the Flight Propulsion Research Laboratory, Cleveland, Ohio

1. Fuel Systems and Controls Laboratory
2. Jet Propulsion Static Test Laboratory
3. Engine Torque Stand No. 1
4. Combustion Tunnel
5. Barrel Storage Building
6. Research Equipment Building
7. Trunnion Storage Building
8. Coal Storage
9. Reservoir and Fire Protection Water Supply
10. Technical Service Building
11. Altitude Wind Tunnel Test Stand

12. Flight Research Building
13. Substation "B"
14. Heating Plant
15. Administration Building
16. Exhauster Building
17. Engine Research Building
18. Exhauster Equipment Wing
19. Compressor and Turbine Research Wing
20. Fuels and Lubricants Building
21. Underground Fuel Storage

22. Cooling Tower No. 1
23. Refrigeration Building
24. Altitude Wind Tunnel
25. Air Dryer
26. Icing Research Tunnel
27. Utilities Building
28. High-Pressure Combustion Laboratory
29. 8 X 6-Foot Supersonic Wind Tunnel
30. Fuel Storage Tanks
31. Engine Propeller Research Building No. 2
32. Engine Propeller Research Building No. 1

Figure 2–8. Aerial view of the Flight Propulsion Research Laboratory, Cleveland, Ohio, 1946. (NACA C-15680)

accomplished by passing heated gases through passages within the component walls, by internal or external electrical heating pads, or by electrical eddy currents induced by a pulsating magnetic field in the elements to be protected. The hot-gas bleedback system operated by injecting hot gas into the air stream at the inlet, which protected the engine and most components in the inlet duct from icing. Finally, inertia separation of super-cooled water droplets out of the engine inlet air stream would be accomplished by a special inlet and duct design.

Using scaled mockups of the engine inlet components, the researchers tested the effectiveness of various ice-protection methods in simulated icing conditions with temperatures as low as -30°F and airspeeds up to 300 miles per hour. Measuring cloud droplet size and liquid water content in the tunnel by the rotating cylinder method, they were satisfied that the simulated cloud was "in the range" of natural icing conditions.

Their investigation revealed that only the surface heating and hot-gas bleedback systems afforded adequate ice protection. "Surface heating," the researchers concluded, "either by gas heating or electrical means, appeared to be the most acceptable icing-protection method with regard to performance losses. Hot-gas bleedback, although causing undesirable thrust losses, offers an easy means of obtaining icing protection for some installations. The final choice of an icing-protection system would depend upon the supply of heated gas and electrical power available and on allowable performance and weight penalties associated with each system."[26]

On 26 and 27 June 1947, AERL hosted its first conference on aircraft ice prevention. The meeting had been organized to convey the latest research results to individuals and organizations responsible for the design, development, and flight application of aircraft ice-protection equipment. It was an opportunity to showcase the work that Cleveland's icing division had accomplished over the past three years.[27]

After reports by researchers at Ames on the progress of their work to develop a practical thermal system, Hunter led off the AERL contingent with an optimistic summary of Cleveland's progress with induction system icing protection. It was not possible, he observed, to design an efficient and ice-free system that would not require the pilot to control heating nor apply alcohol for emergency de-icing.[28] Fleming and von Glahn followed with progress reports on this work to protect jet inlets from icing. Gray and Lewis then spoke about their propeller de-icing experiments. Finally, Callaghan reported on engine-cooling fan de-icing tests that had been conducted in the IRT.[29]

While the conference provided an excellent forum for discussion of the icing research that had been conducted at AERL over the past three years, all was not well with the Cleveland experimenters. Despite a substantial amount of effort and expense, the creation of a proper icing cloud in the IRT remained a vexing problem. Abe Silverstein, chief of the Wind Tunnels and Flight Division, was growing increasingly impatient with the situation. Never known for his even temper, Silverstein communicated his concern to Hunter in the clearest possible language.[30]

The problem faced by the icing division had been spelled out in April 1947 in a report by professor H. G. Houghton of the NACA's Subcommittee on De-Icing Problems. "Even with our incomplete knowledge of the properties of natural icing clouds," Houghton had

[26] von Glahn, Callaghan, and Gray, "NACA Investigations of Icing-Protection Systems for Turbojet-Engine Installations," NACA RM E51B12 (1951).

[27] *Wing Tips*, 11 July 1947. Wing Tips was the house organ of the Lewis Research Center; copies are on file in the History Office, GRC.

[28] Hunter, "Summary of Induction-System Ice-Protection Requirements for Reciprocating-Engine Power Plants," NACA Conference on Ice Prevention, pp. 71–85.

[29] NACA Conference on Aircraft Ice Prevention, pp. 86–112, 119–136.

[30] Interview with Vern G. Rollin by William M. Leary, 23 September 2000.

written, "it is obvious that present techniques for simulating icing conditions are entirely inadequate." Researchers needed suitable instruments for measuring the properties of the artificial cloud in the tunnel. Also, there had to be better control of drop-size distribution. The main problem was the production and distribution of a suitable spray. "I am forced to conclude," Houghton had reported, "that it will not be possible to produce uniform icing conditions over more than a small fraction of cross-section of the measuring section of the present icing tunnel at AERL."[31]

Despite Houghton's pessimism, the next year, during which the name of the Cleveland facility was changed to the Lewis Flight Propulsion Laboratory, saw substantial progress in developing a spray nozzle that would produce the needed droplet sizes. Given carte blanche by Silverstein to solve the problem, engineer Glen Hennings put together a team that modified the Hunter-designed nozzle that had been used earlier by Fleming in the Altitude Wind Tunnel tests. Instead of a 0.010-inch diameter hole in the water tube, Hennings substituted a 0.020-inch-diameter hypodermic tubing (internal diameter about 0.010-inch) and soldered it into each nozzle. The resulting droplets could then be atomized by air flowing around the tubing, and much less compressed air was required. In November 1948, Vern G. Rollin, Hunter's deputy, reported, "water spray nozzles capable of producing narrow and wide ranges of droplet size and distribution from 5 to 200 microns have now been developed."[32]

There still remained the problem of producing a uniform 4-foot by 4-foot icing cloud. Some 40 to 50 of the new nozzles were placed on six horizontal spray bars, and test runs began. Although the nozzles were built to the same specifications, they tended to act differently. In addition, some of the water tube holes would periodically plug up due to small particles in the water because the Cleveland water contained numerous minerals. Engineer Christenson solved this problem by purchasing a demineralizer and designing a system that could handle the demineralized water. Still, the air flow in the tunnel was not uniform. As a result, calibration of the new spray system became a time-consuming, often frustrating process. "We worked for months," researcher Thomas F. Gelder recalled, "experimenting with different nozzle locations and with a range of air and water pressure levels." By 1950, however, it had become possible to produce a uniform icing cloud in the tunnel. It had taken over five years, and it represented a significant technological achievement for the Cleveland icing team. The way was now clear to accelerate research efforts in the IRT.[33]

[31] Houghton, "Progress Report for Meeting of Sub-Committee on De-Icing Problems," 9 April 1947, NASA History Office.

[32] Christenson interview; Rollin interview; Rollin, "Progress Report of Lewis Propulsion Research Laboratory," 8 November 1948, NASA History Office.

[33] Christenson interview; interview with Thomas F. Gelder by William M. Leary, 18 June 2000.

Chapter 3

A Golden Age

While work on perfecting the spray system in the IRT was taking place, Cleveland's role as the center for the NACA's icing research was enhanced with the decision to close the ice-prevention program at the Ames Research Center and transfer flight research to AERL/Lewis. Ever since its opening in October 1940, Ames had led the way in developing a practical thermal de-icing system. Emphasizing flight research, the experimenters at Ames, under the driving leadership of Lewis Rodert, had flight-tested various thermal de- and anti-icing designs. The results had led to the development of a complete thermal system, covering the wings, tail surfaces, engine nacelles, windscreen, and other vulnerable parts of the airplane. In the spring of 1948, however, NACA Headquarters decided to consolidate icing research. As it was far easier to move the aircraft than the IRT, the C-46 and B-24 that were being used for flight tests at Ames were transferred to Cleveland, together with meteorologist William Lewis of the Weather Bureau.[1]

The NACA's flight research program never missed a beat with the move to the Midwest. One area of continuing interest concerned the physics of icing clouds. Researchers wanted to know how liquid water content and droplet size varied with temperature, pressure, geographic location, synoptic conditions, and topography. The work on this subject that was begun at Ames continued at Cleveland. Numerous flights were made over Lake Erie during which ice was collected on cylinders of different diameters that rotated on a common axis and were exposed outside the thermal-protected C-46. The multicylinders were retracted after a flight through icing conditions, disassembled, and the ice was weighed. Comparing the weight of the ice with calculated values of collection efficiency (obtained theoretically) produced data on liquid water content, average droplet size, and droplet-size distribution. Although the data obtained from the rotating cylinder collection system was reliable, the

---

[1] Bugos, "Rodert;" Sharp to NACA Headquarters, 25 May 1948, History Office, GRC.

method was cumbersome, and it was applicable only in clouds where the temperature was below 32°F.[2]

The need for additional data led the Cleveland researchers to develop an automatic icing rate meter. This instrument operated on a differential pressure basis. When tiny orifices in the device were plugged by ice, a pressure switch was activated that turned on heat to eliminate the ice formation. When the ice was gone, the change in pressure automatically turned off the heat. From the duration of the cycle, researchers could determine the rate of ice accumulation. By the winter of 1950–51, the perfected device was being carried onboard aircraft of Northwest, United, American, and TWA on routes throughout the continental United States, Alaska, and across the North Atlantic. The U.S. Air Force also participated in this program, providing worldwide data.[3]

NACA researchers discovered that the average droplet size in icing clouds was in the range of 10 to 25 microns in diameter. Maximum water content was about 1.5 grams per cubic meter in stratus clouds and as high as 3.5 grams in cumulus clouds. High water content, however, did not extend for more than 0.5 of a mile in cumulus clouds and 10 to 20 miles in stratus clouds.[4]

This information not only assisted icing researchers in their work, but also provided the basis for ice-protection design standards that later were adopted by the Civil Aeronautics Administration. Civil Air Regulations (CARs) before 1953 required only that if de-icing boots were installed, there must be a positive means of deflating all wing boots. In December 1953, Part 4b of the CARs expanded these requirements to include cockpit vision in icing conditions, a heated pitot tube for airspeed indication, propeller de-icing, protection of induction systems, and other anti-icing and de-icing requirements.

A major addition to the regulations took place in August 1955 with the introduction of icing envelopes. Using mostly the data that NACA researchers had accumulated during multicylinder flights, the CAA defined icing envelopes in terms of liquid water

---

[2] R. J. Brun, W. Lewis, P. J. Perkins, and J. S. Serafini, "Impingement of Cloud Droplets and Procedure for Measuring Liquid-Water Content and Droplet Size in Supercooled Clouds by Rotating Multicylinder Method," NACA Report 1215 (September 1955); Perkins interview.

[3] P. J. Perkins, S. McCullough, and R. D. Lewis, "A Simplified Instrument for Recording and Indicating Frequency and Intensity of Icing Conditions Encountered in Flight," NACA RM E51E16 (1951); *Wing Tips*, 24 November 1950; William Lewis, "Icing Conditions to be Expected in the Operation of High-Speed, High-Altitude Airplanes," NACA Conference on Some Problems of Aircraft Operation, 17–18 November 1954.

[4] Alun R. Jones and William Lewis, "Recommended Values of Meteorological Factors to be Considered in the Design of Aircraft Ice-Prevention Systems," NACA TN 1855 (1949).

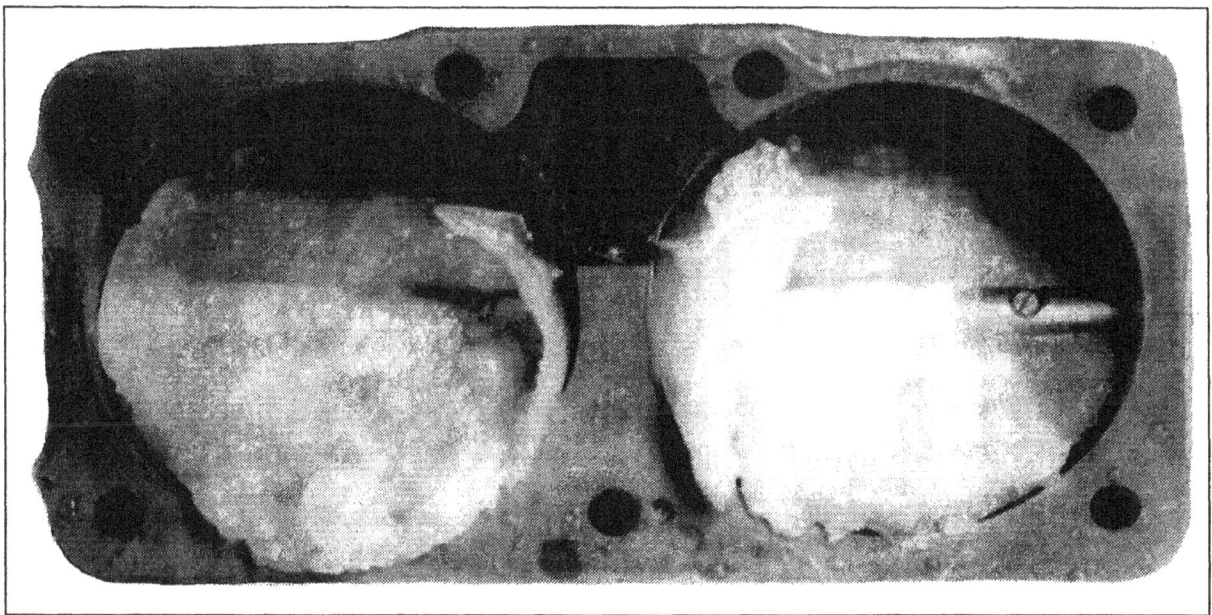

Figure 3–1. Ice formed on the butterfly valves of an aircraft engine carburetor during tests in the IRT, 28 July 1944. (NACA C-5690)

content, mean effective diameter of droplets, temperature, and horizontal and vertical extent of the super-cooled icing cloud environment.[5]

When Part 4b was re-codified into Part 25 in February 1965, these icing envelopes were incorporated into the new CAR as Appendix C. Part 25 provided that transport aircraft must be able to operate safely through forecasted continuous and intermittent maximum icing conditions as determined by the icing envelopes in Appendix C. It further specified that the effectiveness of ice-protection systems had to be shown by flight tests in natural and simulated conditions and by icing tunnel tests.[6]

The flight research into thermal systems that had begun at Ames applied primarily to reciprocating-engine aircraft. Until 1950, with the exception of the research on jet engine inlet protection, work in the IRT also had been weighed heavily in the direction of piston-engine airplanes. But the NACA was well aware that the successful flight of the first turbojet aircraft on 14 May 1941 had marked the beginning of a new era in the history

---

[5] W. H. Weeks, CAA, to R. V. Rhode, NACA, 15 April 1954, NASA History Office, discussed the change in regulations.

[6] D. T. Bowden, A. E. Gensemer, and C. A. Skeen, "Engineering Summary of Airframe Technical Icing Data," Technical Report ADS-4, Federal Aviation Agency Contract FA-WA-4250 (March 1964). ADS-4 was considered to be the "Icing Bible" until the appearance of the *Aircraft Icing Handbook* in 1991.

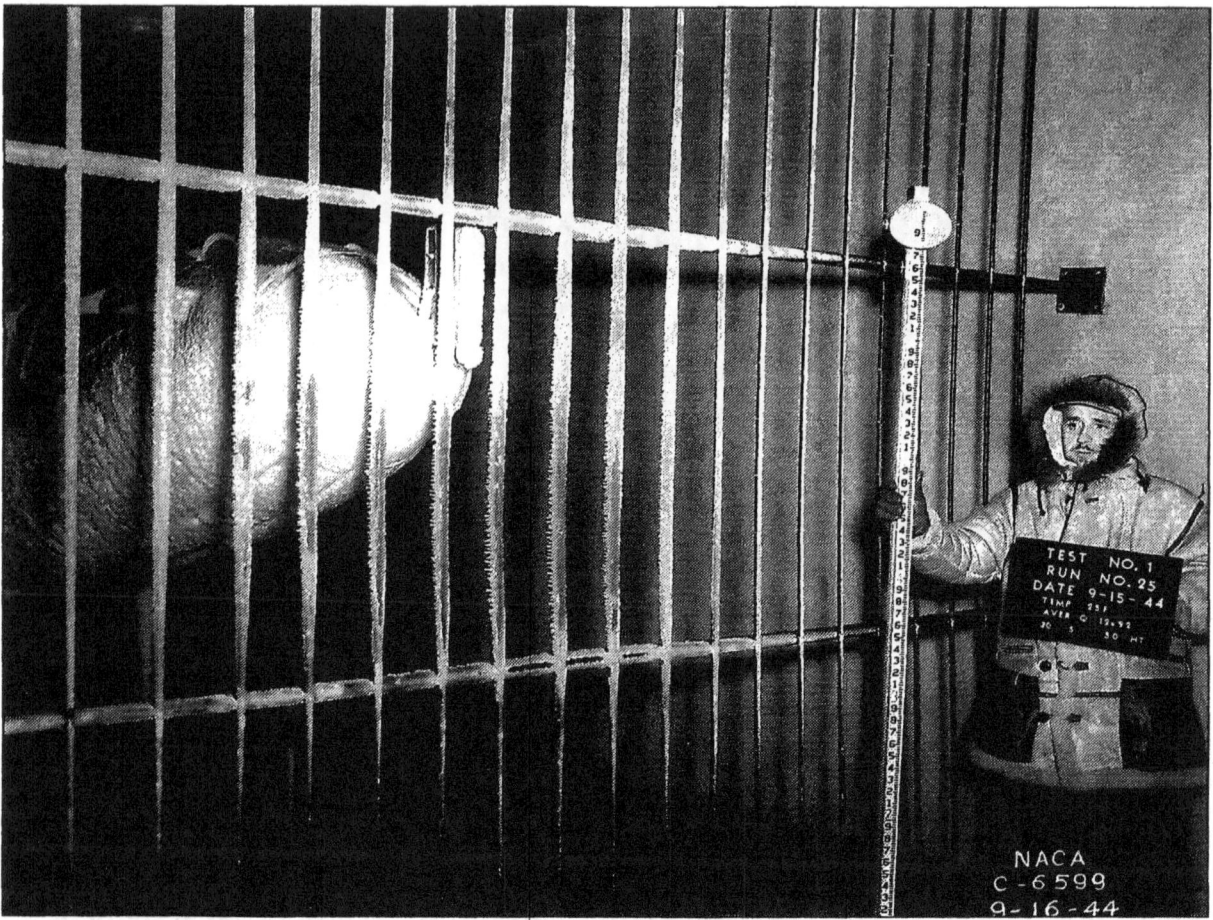

Figure 3–2. Propellor with internal heaters after icing run, 15 September 1944. (NACA C-6599)

of aviation. Thin-winged interceptor aircraft, the NACA concluded, with their high rates of climb and descent and high cruising altitudes, might not require airframe icing protection. Jet transports, on the other hand, would need such protection during climb and descent. Both interceptors and transports had to have complete engine protection.[7]

By the early 1950s, Cleveland's focus had shifted to jet aircraft icing protection. Tunnel usage shot up dramatically, as researchers sought solutions to the icing problems of high-altitude, high-speed aircraft. At the beginning of the decade, researchers worked in the IRT about 500 hours a year. In 1951, however, usage increased to 651 hours. The IRT operated for 917 hours in 1953, 871 hours in 1954, and 817 hours in 1955. Peak

---

[7] von Glahn, "Some Considerations of the Need for Icing Protection of High-Speed High-Altitude Airplanes," NACA Conference on Some Problems of Aircraft Operation.

Figure 3–3. James P. Lewis inspects fuselage of a P-39 mounted in IRT for study of ice formation on propellors. (NACA C-10186)

staffing of the facility came in June 1954 with thirteen professionals, six other personnel, and twelve supporting personnel.[8]

Management of the branch also changed, as von Glahn replaced Hunter, who became chief of research at Lewis.[9] Born in Germany in 1919, von Glahn's family had moved to the

---

[8] C. S. Moore, "Research Facilities Operation Summary, 1944–1955," n.d., History Office, GRC.

[9] The circumstances and exact date of Hunter's departure are uncertain, although the change in leadership appeared to take place in late 1950 or early 1951. In the early 1960s, Hunter became director of conferences in NASA's Public Affairs Office in Washington, DC. In 1966, he was appointed NASA's senior scientific representative to Australia. He retired in 1979 and died in 1983. The information on his career is taken from the obituary clipping in the NASA History Office.

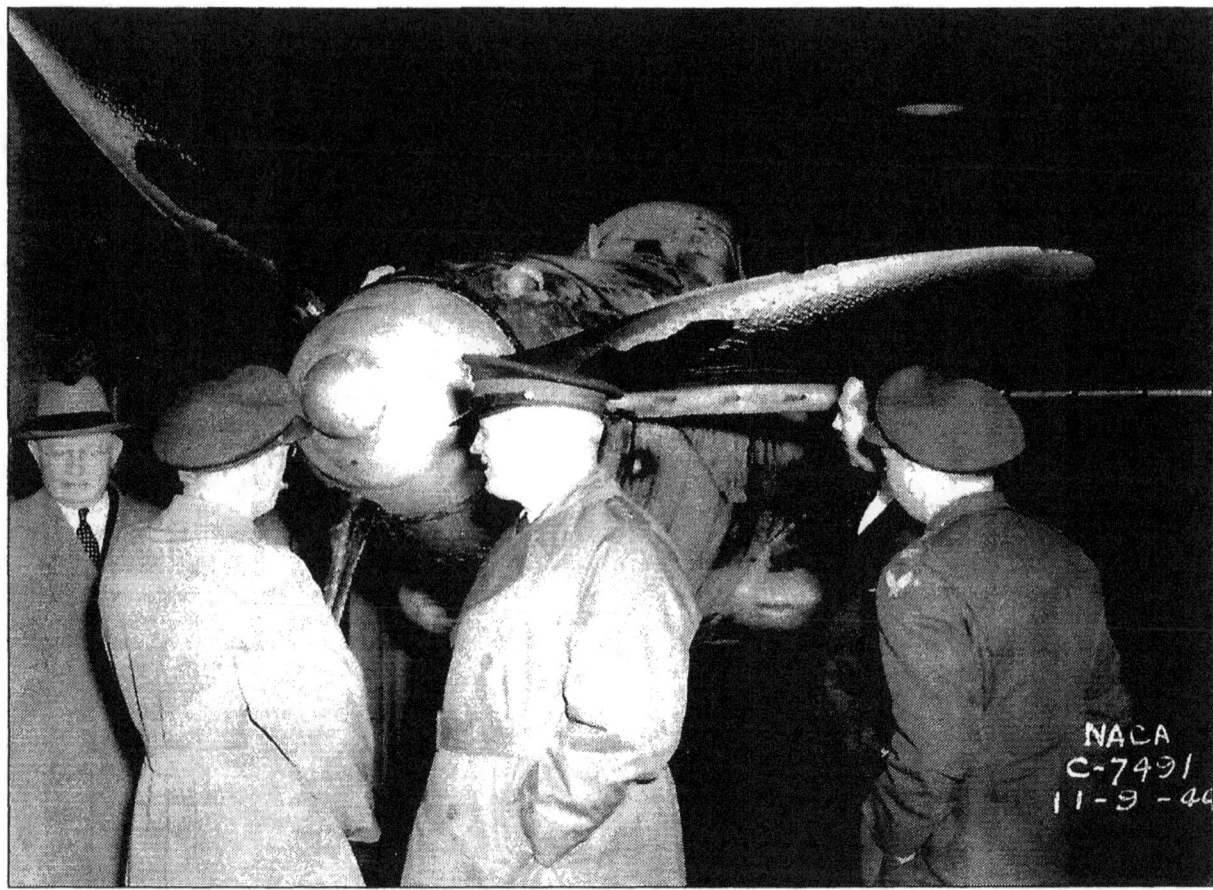

Figure 3–4. Gerald H. H. Arnold views an icing experiment during a visit to AERL, 9 November 1944. Dr. George Lewis is on the far left; Willson Hunter is between Arnold; and the Colonel is on the right. (NACA C-7491)

United States in 1926 due to the instability of the Weimar Republic. Von Glahn was raised in Pennsylvania, New Jersey, and New York. In 1942, he graduated from Rensselaer Polytechnic Institute with a degree in aeronautical engineering. He remained at RPI until 1944 as an instructor in the aeronautics department, teaching and doing wind tunnel research. Gordon Campbell of BFGoodrich, who also was an RPI alumnus, told von Glahn about a position in icing research that was opening at the new NACA laboratory in Cleveland. Von Glahn was hired by NACA just prior to the beginnings of IRT operations.[10]

In designing ice-protection systems for large jet transports, manufacturers first needed information on what happened when super-cooled water droplets struck an air-

---

[10] von Glahn interview.

Figure 3–5. Outside view of the IRT, c. 1945. (NACA C-10763)

foil. The NACA responded with a series of theoretical and experimental droplet trajectory studies. The Cleveland researchers came up with the idea of employing a water-soluble dye in the IRT to study impingement patterns. Von Glahn's father, Dr. William von Glahn, who was in charge of dyestuff research at GAF's chemical plant in Rensselaer, New York, suggested that a water-soluble Azo Rubine dye might be suitable. He was correct. Thomas F. Gelder, who conducted most of the experiments, added the red dye in low concentrations (about 1 percent) to 5-gallon Pyrex jugs of water, which he then placed on a hot plate and heated to promote mixture. Usually, this was a routine procedure. In one instance, however, a jug on the hot plate broke and dumped dyed water down the drain and into Rocky River, which emptied into Lake Erie. The red water caused a good deal of comment in the community.[11]

---

[11] von Glahn interview; Gelder interview.

Figure 3–6. Rig for testing propellor ice-protection systems, 1946. (NACA C-15254)

The dyed water was placed in 20-gallon tanks, which were connected to six spray nozzles in the IRT. Gelder then taped blotting paper to the airfoil that was to be tested. After the tunnel was operating, a 5- to 10-second spray of water was released into the airstream. The blotter strip was then removed, and a special punch was used to cut out rectangular strips, 0.125 of an inch wide by 1.5 inches long. Each strip was placed in a numbered test tube on a long rack, and distilled water was added. A light beam passing through the mixture would determine the dye concentration in the water.[12]

The dye impingement research involved a good deal of tedious work for the researchers, and, later, the women who usually did the required hand calculations were brought in to handle the task. Jane Gavlak recalls that Gelder's strips had to be placed in boiling water to remove the dye. She then had to match the resultant dye against a comparison chart to determine concentration. Although the work was hard on both eyes and hands, the results were worth the effort. For the first time, the researchers could determine where the droplets were striking and in what concentration. Thus, the principal problem areas that needed icing protection could be identified.[13]

Although the trajectory studies provided information on the rate that ice would accumulate on an airfoil, NACA researchers still needed reliable data on the shape of the ice formations. Initially, flight tests had been made to measure the aerodynamic penalties caused by various ice shapes. Researchers found that rime icing, primarily associated with low temperatures, had a streamlined shape and did not greatly affect performance. However, heavy glaze icing, associated with high water content and temperatures near freezing, was a different story. Glaze ice formations protruded into the airstream and caused significant aerodynamic penalties.[14]

The controlled conditions of the IRT allowed researchers to test airfoils of different sizes, thicknesses, and shapes over a wide range of angles of attack and icing conditions. Work done in the early 1950s by von Glahn and Vernon Gray produced drag curves for NACA 63A-009 (6.9-foot chord, 36-inch sweep) and NACA 65-212 (8-foot chord) air-

---

[12] Gelder interview; Gelder, W. H. Smyers, and von Glahn, "Experimental Droplet Impingement on Several Two-Dimensional Airfoils with Thickness Ratios of 6 to 16 Percent," NACA TN 3839 (1956).

[13] von Glahn, Gelder, and W. H. Smyers, "A Dye-Tracer Technique for Experimentally Obtaining Impingement Characteristics of Arbitrary Bodies and Method for Determining Droplet Size Distribution," NACA TN 3338 (1955); Leary interview with Jane Gavlak Zager, 27 June 2001.

[14] For an excellent summary of the NACA's icing research during this period, see von Glahn, "The Icing Problem - Current Status of NACA Techniques and Research," originally presented at the Ottawa AGARD Conference, 10–17 June 1955, and reproduced in "Select Bibliography of NACA-NASA Aircraft Icing Publications," NASA TM 81651 (1981), pp. 1–10, together with microfiche copies of all important technical papers.

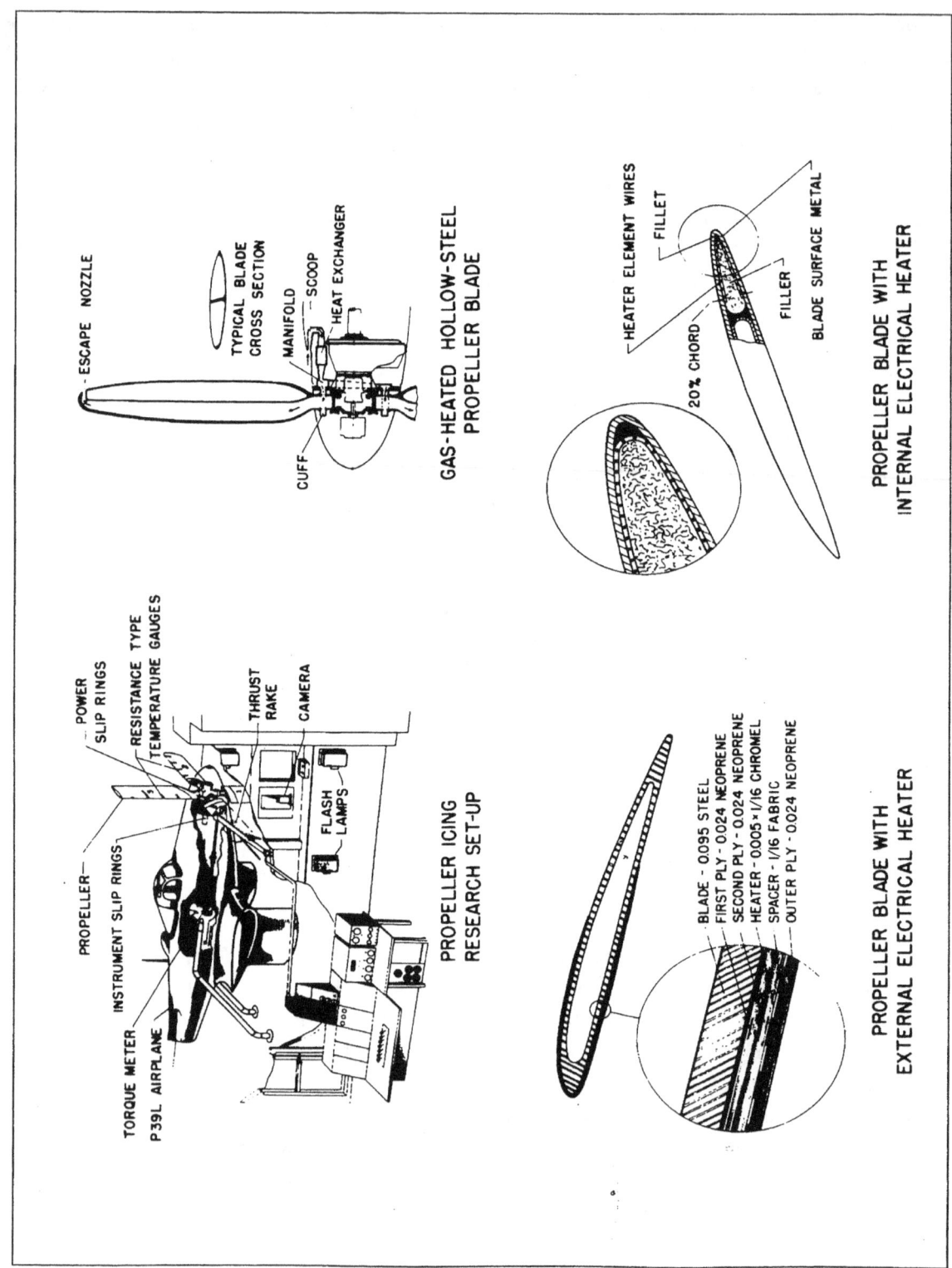

Figure 3–7. Icing Research Tunnel investigation of the effectiveness of various propellor anti-icing devices, 1946. (NACA C–15557)

foils under a variety of icing conditions. For example, at high angles of attack (8 °), ice formations on the upper surface near the leading edge of the NACA 65-212 airfoil caused large increases in drag and incipient stalling.[15]

The data points for these—and all other—experiments were obtained with the assistance of a corps of young women who were known as "computers." During experiments, researchers would take about 100 data points during an evening's test run. A bank of some 100 manometer tubes, 0.75 of an inch apart and with a 0.1-inch scale between the two tubes, recorded pressure, strain gauge readings, and other information. Cameras would take photographs of the manometers at different stages of the experiment. The next day, a stack of these photographs would appear in the NACA's computing section, located on the second floor of the 8 x 6-foot tunnel building. There were three offices in the section: a "big office" which contained some thirty-five young women; a middle office, with twenty to twenty-five women; and a small office, with ten women with advanced mathematical skills. Jane Gavlak, who worked in the "big office" between 1948 and 1954, was one of the "computers" who would read and plot the measurements from the tubes. Reading and plotting one manometer board of 100 tubes would yield 1 data point. As the experiments would take 100 data points in an evening, the researchers would not get the readings for two or three weeks.

It took "hours and hours" of tedious work to read and plot the results, Gavlak remembered. Nonetheless, it was a good job for a high school graduate. It paid well ($1,900 a year for a 40-hour week), and there was little turnover in the section. Women usually left only for marriage and pregnancy. Morale was high, as researchers often would involve the women in their work. An experienced "computer" could often identify problems with the data and alert researchers that there might have been a mistake in taking the information. Such an alert could save weeks of efforts for a researcher.[16]

During the 1940s, NACA research had established the feasibility and design basis for a thermal anti-icing system that continuously applied sufficient heat to critical aircraft components to maintain impinging super-cooled droplets in a liquid state over the entire surface. This was the research, based on studies of propeller-driven aircraft, that had won Rodert the Collier Trophy for 1946. In 1950, however, Gelder and Lewis presented a paper at a meeting of the Institute of Aeronautical Sciences (the predecessor of the American Institute of Aeronautics and Astronautics) that called into question the via-

---

[15] von Glahn and Gray, "Effect of Ice Formations on Section Drag of Swept NACA 63A-009 Airfoil with Partial Span Leading Edge Slat for Various Modes of Thermal Ice Protection," NACA RM E53J30 (1954); Gray and von Glahn, "Effect of Ice and Frost Formations on Drag of NACA 65-212 Airfoil for Various Modes of Thermal Ice Protection," NACA TN 2962 (1953).

[16] Jane Gavlak Zager interview.

Figure 3–8. First icing conference at Lewis, 26 June 1947. (NACA C-19051)

bility of a continuous heat system for large high-speed, high-altitude jet aircraft. Heating requirements for jet transports, they found, could be enormous—up to 7,500,000 Btu per hour under certain conditions, or 10 percent bleed of the engine airflow. Performance penalties and excessive fuel consumption of this magnitude could preclude the use of continuous heating on jet aircraft.[17]

The alternative to continuous heating would be cyclic de-icing. In a cyclic system, ice would be allowed to accrete on an airfoil for a short period of time. A short but intense application of heat would then be applied to melt the bond of ice to the surface of the airfoil and allow it to be removed by aerodynamic forces. The heat would be ter-

---

[17] The 1950 paper evolved into Gelder, Lewis, and Stanley L. Koutz, "Icing Protection for a Turbojet Transport Airplane: Heating Requirements, Methods of Protection, and Performance Penalties," NACA TN 2866 (1953).

minated and the surface allowed to cool and re-ice before another application of heat would be necessary. Although cyclic de-icing previously had been studied for propellers and jet engine guide vanes, the results of these investigations did not apply to the airfoils that were found on the wings and empennages of jet aircraft. In order to obtain information on cyclic de-icing systems that could be used by jet transports, the NACA launched a major investigation of the various types of heaters and methods of heating in the early 1950s.

The investigation began with an examination of the characteristics and requirements of de-icing a NACA 62 (2)-216 airfoil by using an external electric heater. Researchers Lewis and Bowden mounted the airfoil, which had an 8-foot chord and a 6-foot span, vertically in the test section of the IRT. Technicians installed an external electric heater on the forward section of the model, extending chordwise a distance of 14.1 percent chord on the upper surface and 23.4 percent cord on the lower surface. The heater consisted of 0.125-inch-wide Nichrome resistance strips, each 0.001-inch thick and spaced 0.0313 of an inch apart. The heating ribbons, enclosed between two layers of neoprene, were connected to a variable cycle timer that allowed researchers to control the heat-on and heat-off periods. A recording wattmeter measured total power input to each heater element.

The airfoil also featured a parting strip to facilitate ice removal. The parting strip consisted of a 1-inch-wide spanwise area located near the airfoil stagnation region that was continuously heated. During icing tests, researchers found that quick and complete ice removal, when cyclic heat was applied to the airfoil, could only be achieved with the assistance of the parting strip.

In conducting the icing tests, Lewis and Bowden used a variety of air temperatures, airspeeds, angles of attack, droplet sizes, and liquid water content. Their aim was to determine the minimum power output that was required for complete and consistent ice removal. They found that the most important variables in determining power requirements were temperature and heat-on time. Heat-off time, droplet size, and liquid water content had only a secondary effect, while angle of attack had no appreciable effect. High local application of power and short—less than 15 seconds—heating periods provided the most efficient removal of ice, with a maximum total energy output of only 490,000 Btu per hour.[18]

While an electric de-icing system accomplished the task of removing ice with low energy outputs, it had numerous disadvantages. It added to the weight of the aircraft, was susceptible to failure by damage to the heating circuits, was costly to maintain, and posed

---

[18] Lewis and Bowden, "Preliminary Investigation of Cyclic De-Icing of an Airfoil Using an External Electric Heater," NACA RM E51J30 (1952).

Figure 3–9. Water spray test in the IRT, 22 September 1947. (NACA C-19747)

a potential fire hazard if the system failed because of heater burnout. The alternative was a hot-gas system. If the large heating requirements associated with continuous hot-gas heating could be reduced by a cyclical system, ice could be removed without an increase in aircraft weight. A hot-gas cyclical system also had the advantages of integral design with the aircraft structure, low maintenance costs, and no fire hazards.

For tests of the hot-gas cyclical de-icing system, researchers Gray, Bowden, and von Glahn used a NACA 65(1)-212 airfoil of 8-foot chord. The leading edge of the airfoil was heated by gas flow through chordwise passages in a double-skin construction similar to that found in a continuous gas-heating system. The airfoil also incorporated the parting strip that had been employed in the electric de-icing system experiments.

The researchers subjected the model to a wide range of conditions. They used air-speeds of 180 and 280 miles per hour, angles of attack from 2° to 8°, air temperatures of -11° to 20°F, and liquid water content of 0.3 to 1.2 grams per cubic meter. To melt the ice, they employed high-pressure heated air, regulating the temperature by adding cold air, and controlling the pressure by means of a pressure-regulating valve. Gas temperature at the inlet of the supply duct ranged from 200° to 510°F.

Their investigation revealed that the ice could be removed satisfactorily with cycle ratios (total cycle time divided by heat-on periods) of 10 to 26. "For minimum runback, efficient ice removal, and minimum total heat input," they concluded, "short heat-on periods of about 15 seconds with heat-off periods of 260 seconds gave the best results." Savings in heat over continuous anti-icing system requirements were considerable.[19]

The final phase of the investigation involved a comparative study of several methods of cyclic gas-heating systems. Researchers Gray and Bowden again used as their model a NACA 65(1)-212 airfoil with a 6-foot span, 8-foot chord, and a maximum thickness of 11.5 inches. They tested three systems that differed mainly in the way that they obtained elevated gas temperatures at the leading edge and in their use of parting strips. One system used a double-duct return-flow gas supply arrangement with spanwise and chord-wise parting strips. A second system was similar to the first one but without the parting strips. The third system featured a single-duct non-return gas supply arrangement and no parting strips.

The first system gave the best results. Gray and Bowden found that 50 percent longer heat-on periods were required for systems without parting strips. They also discovered that the single-passage gas-supply duct system needed an 85 percent longer heat-on period than did the dual-duct systems. Overall, heat source requirements for

---

[19] Gray, Bowden, and von Glahn, "Preliminary Results of Cyclical De-Icing of a Gas-Heated Airfoil," NACA RM E51J29 (1952).

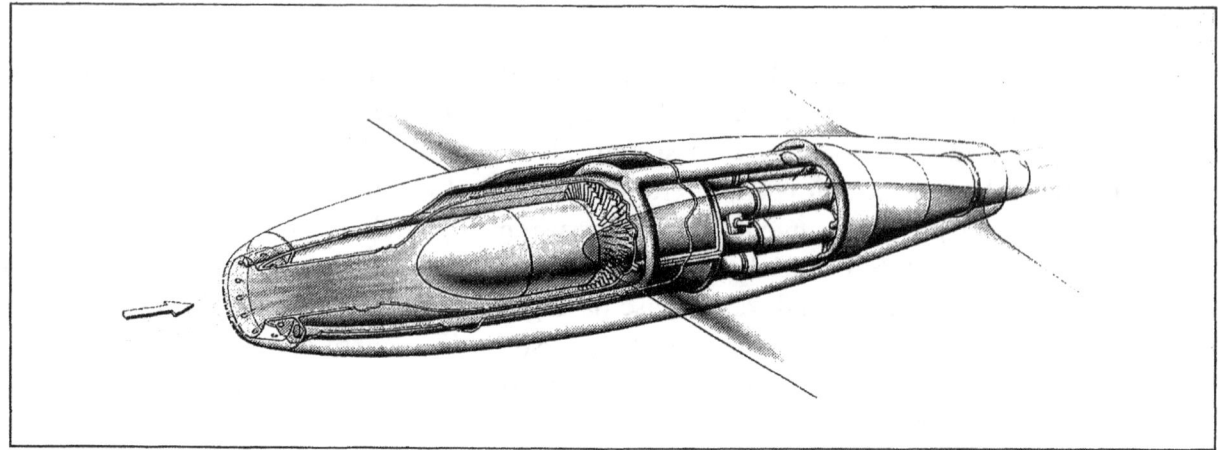

Figure 3–10. Turbojet ice prevention with hot-gas bleedback, 1947. (NACA C-19864)

cyclic de-icing were between one-fourth and one-tenth of those used by comparable con-tinuous gas-heated anti-icing systems.[20]

Work also continued at Cleveland on systems to protect the engines of jet aircraft. The hazards of this type of icing were dramatically demonstrated in 1951 when eight U.S. Air Force F-84 fighter-bombers were lost over Richmond, Indiana, on 8 June during a brief encounter with severe icing conditions. The aircraft were part of a flight of thirty-four Thunderjets that had just taken off from Wright-Patterson Field, Ohio, for Selfridge Field, Michigan. While flying through a thunderstorm, eight of the aircraft experienced sudden engine failure. Three pilots died and two were injured in the largest mass jet disaster in aviation history. At first, the military suspected sabotage and called in the Federal Bureau of Investigation. But it soon became clear that icing was to blame. Ice had built up on the engine inlet screens that prevented the ingestion of debris during takeoff, choking off the airflow and causing the engines to quit. The Air Force turned to the NACA for an answer to this problem. Tests in the IRT, however, revealed that the screens could only be protected with an unacceptably large heating requirement. The NACA recommended that the screens be removed or be made retractable.[21]

Manufacturers also came to the NACA with their problems. In December 1954, for example, Convair requested the NACA to conduct icing tests of the inlet for the

---

[20] Gray and Bowden, "Comparisons of Several Methods of Cyclic De-Icing of a Gas-Heated Airfoil," NACA RM E53C27 (1953).

[21] *New York Times*, 9, 10, and 13 June 1951; von Glahn, "The Icing Problem;" von Glahn, Edmund C. Callaghan, and Vernon H. Gray, "NACA Investigation of Icing-Protection Systems for Turbojet-Engine Installations," NACA RM E51-B12 (May 1951).

General Electric J-79-GE-1 turbojet engine that would power their advanced super-sonic B-58 Hustler. The stainless-steel cowl of the engine was heated by internal electric heating elements manufactured by BFGoodrich, while a gas-heat system that was designed and manufactured by Convair protected the center body and support struts. The NACA's plan called for researchers to secure aerodynamic and droplet impingement data, which would be followed by tests of the ice-prevention systems for inlet components.

Convair provided two inlet models for tests in the IRT—one unheated and one heated. The unheated model arrived in April 1955. The research scheduled called for twelve aerodynamic runs, followed by nine dye runs, with an average time of 6 hours per run. Icing studies with the heated model would take place from mid-June to mid-July.

While the dry air aerodynamic and droplet impingement studies were satisfactorily concluded, problems showed up during the icing runs. The gas-heated center body and support structure performed well in the icing tests, but the electrically heated cowl proved unsatisfactory. Researchers found that the cowl lip could not be protected completely at air temperatures below 25°F because the heater installation left a cold area at the cowl leading edge. As an alternative to the electric-heat system, the NACA suggested that Convair adopt a gas-heated cowl.[22]

In addition to airframes and engines, radome ice protection also attracted the attention of NACA researchers. By the early 1950s, radar had become an important component of commercial transports and all-weather military aircraft. U.S. Air Force interceptors employed radar not only for weather information, but also for target tracking and fire control. In order to evaluate the icing and icing protection of radomes, two Northrop F-89 Scorpion domes were tested in the IRT. The APG-33 radar of the F-89C was housed in a narrow, parabolic dome with a nose radius of 6.955 inches, while the larger APG-40 radar used by F-89Ds occupied a blunt, hemispherical dome that had a nose radius of 13.91 inches. Both radomes (0.375-inch thick) were constructed of molded fiberglass that had been impregnated with synthetic resin and coated with a rubber-like material to resist erosion and abrasion.

Researchers James P. Lewis and Robert J. Blade began their three-phase investigation by determining the rate and location of water-droplet impingement, and the manner in which the radome accreted ice. Using a dye-tracing technique, they wrapped strips of absorbent paper around the radome surface, which they then exposed to a water-dye spray solution for 1–10 seconds. A spectrophotometer determined the quantity of dye collected in the strips. As they knew the dye concentration in the spray cloud,

[22] Silverstein to Convair, 7 January 1955, and 2 April 1955; von Glahn to Silverstein, 22 August 1955; all in History Office, GRC.

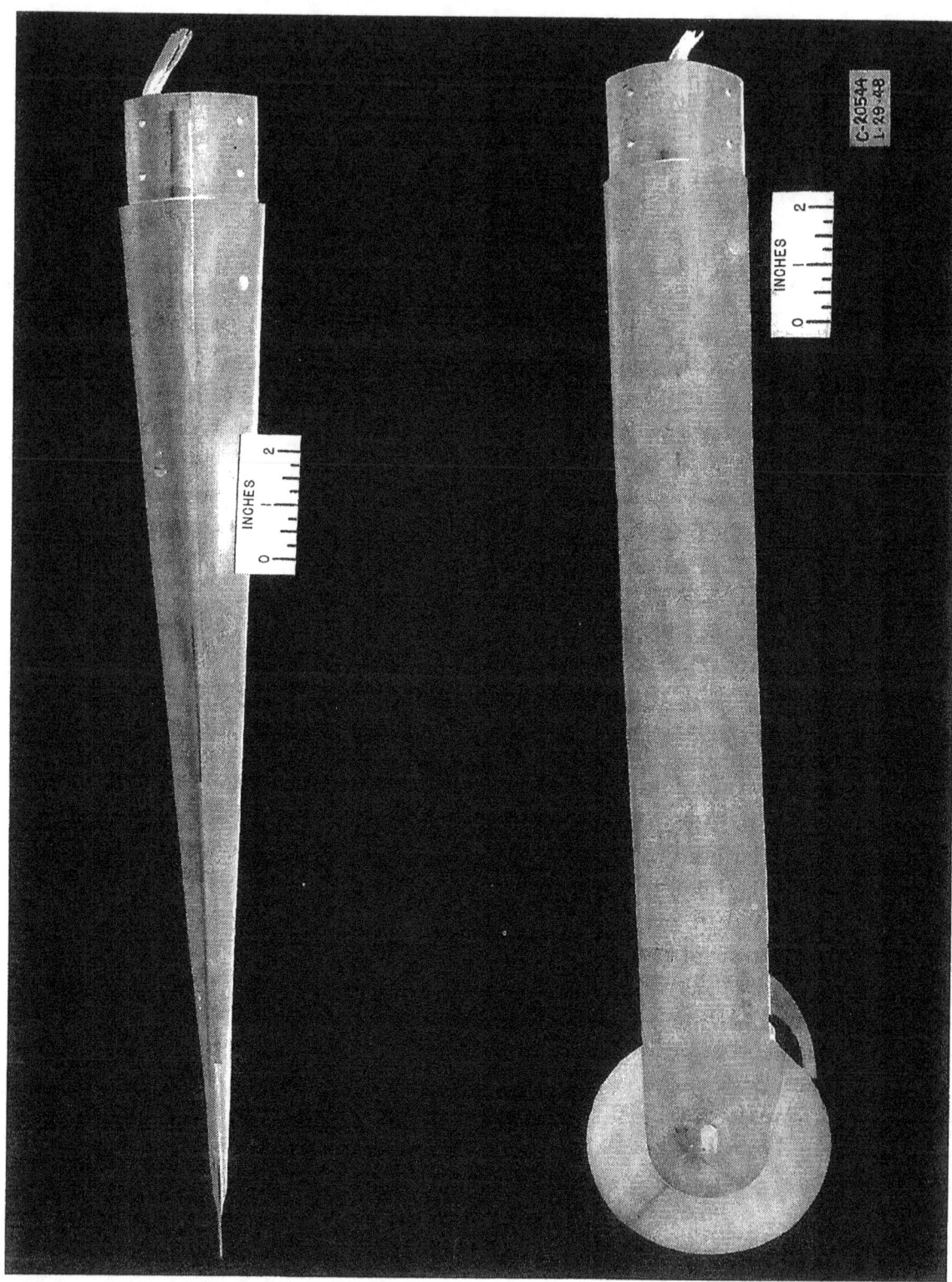

Figure 3–11. Rotating disk icing rate meter, 1948. (NACA C-20544)

they could establish the local rate of deposition of water on the radome and the local collection efficiency.

After acquiring the impingement information, Lewis and Blade next subjected the radomes for 15 minutes to an icing cloud with air temperatures of -5° to 12°F, a liquid water content of 0.9 grams per cubic meter, and droplet diameters of 10 to 12 microns. They conducted the tests at airspeeds of 168 and 275 miles per hour and with angles of attack of 0° and 4°. At the end of the exposure period, they took photographs of the ice formations and measured the ice thickness.

The second phase of the program involved tests of the interference to the radar signals that were caused by icing. Lewis and Blade first tuned the radar system for 100 percent transmission, and then they checked the values against previously obtained measurements in dry-air static tests. After icing the radome for 15 minutes, the researchers again measured transmissions at various scan angles.

In the final phase of the radome experiments, Lewis and Blade assessed the effectiveness of the F-89's fluid-protection system. The system consisted of a spray nozzle located in the radome nose that sprayed a mixture of ethylene glycol and water ahead of the radome. The mixture was swept back on the radome surface by the airstream to remove or prevent icing. The researchers tested several types of air-atomizing and fluid-atomizing nozzles in various spray locations.

The radome investigation determined that the extent of impingement covered most of the radome area scanned by the radar beam in normal operations. Ice formations were typical for similar shapes, while the icing rate varied from 1.25 to 3 inches per hour under test conditions. Radome icing, they found, had a serious effect on the operation of the radar, with significant losses in transmission with ice formations as small as 0.125-inch thick and for an icing period of less than 5 minutes. Provided that a deflecting type of air-atomizing nozzle was used, the fluid-protection system would afford satisfactory de-icing of the radomes with flow rates of less than 4 gallons an hour and for de-icing periods of less than 5 minutes.

As little had been known previously about radome icing, especially the effect of icing on radar transmissions, the NACA study made an especially significant contribution to an area of growing interest by both military and commercial operators.[23]

High-speed, high-altitude jet interceptors generally operated at altitudes of about 20,000 feet, where icing encounters were rare. The aircraft, however, were exposed to icing during the climb and descent phases of flight. In order to obtain data on the frequency and severity of these icing encounters, the NACA and the U.S. Air Force

---

[23] Lewis and Blade, "Experimental Investigation of Radome Icing and Icing Protection," NACA RM E52J31 (1953).

Figure 3–12. Icing rate meter installed on an aircraft, c. 1949. (NACA C-24994)

conducted a one-year program during which interceptors were equipped with pressure-type icing rate meters. Five Air Defense Command F-89s at Duluth, Minnesota, where polar air masses predominated, would carry NACA-designed icing rate meters, mounted near the nose and 7 inches out from the fuselage, from August 1955 to June 1956. Another five F-89s, flying from Seattle, Washington, would sample icing conditions in the maritime air masses of the Pacific Northwest from November 1955 to September 1956.

Researcher Porter Perkins, who analyzed the data, found that the F-89s encountered icing fifty-nine times in 1,178 flights in both areas. In Duluth, 70 percent of these icing incidents took place below 10,000 feet, while in Seattle, 75 percent occurred below 15,000 feet. Ice thickness on the small sensing probe averaged less than 0.0313 of an inch and did not exceed 0.5 of an inch during steep angles of flight for short periods through generally thin cloud layers. Although Perkins cautioned that the limited information from the surveyed areas could be used only as "a very rough estimate of the probabilities of occurrence for icing for other locations and periods," the study did add to the growing body of data that suggested that icing protection equipment might not be necessary for interceptors.[24]

While interceptors could expect only brief exposures to icing, little information was available on the performance penalty that even small amounts of ice might exact. The NACA's research on the aerodynamic effects of icing had been conducted on airfoil sections that were used by large transports and bombers. The high-speed, high-altitude interceptors, however, featured a thin airfoil with a thickness ratio of the order of 4 percent. Clearly, a study of this type of airfoil would be necessary in order for manufacturers and operators to assess the need for icing protection equipment on lifting and control surfaces for high-speed interceptors.

Researchers Gray and von Glahn set out to investigate this problem, using as their model a NACA 65A004 airfoil section of 6-foot chord. They equipped the stainless steel model with a 42-inch-span removable wooden leading-edge section, extending to 27 percent of the chord, that would enable researchers to install a variety of ice-protection systems. They attached the airfoil section to a balance frame in the test section. The frame was connected to a six-component force-balance system. Lift, drag, and pitching moment would be recorded simultaneously on tape by an electrically controlled printing mechanism at each balance scale.

The researchers tested the airfoil at airspeeds of 109 and 240 miles per hour, air temperatures of 0°, 10°, and 25°F, and angles of attack of 0° to 12°. The model was allowed

[24] Perkins, "Icing Frequencies Experienced During Climb and Descent by Fighter-Interceptor Aircraft," NACA TN 4314 (1958).

to collect ice for a period of 3 to 17 minutes, while data was recorded at intervals of 0.5–2 minutes. Gray and von Glahn used an icing spray with a liquid water content of 0.45 to 2.0 grams per cubic meter, as measured with a pressure-type icing rate meter. Droplet size ranged from 11 to 19 microns. The researchers could adjust droplet size by changing the spray nozzle pressure settings. These settings had been determined from a previous calibration of droplet size obtained from tests with water droplets that carried a dye solution. At the end of the icing period, the researchers took photographs to record the shape and size of the ice formations.

Following the lift, drag, and pitching moment studies, technicians removed the trailing edge of the airfoil at the 82 percent chord and modified it to incorporate a simple hinged flap. Researchers then conducted additional icing tests with angles of attack from 3.3° to 8.8° and flap angles up to +/- 15°. Ice formations on the leading edge, they observed, produced no significant changes on the control-surface hinge movements.

An analysis of the aerodynamic data and ice formation photographs showed that the magnitude of aerodynamic penalties was primarily a function of the shape and size of the ice formations near the leading edge of the airfoil. And the shape and size of these ice formations, Gray and von Glahn noted, were functions "of such variables as water content, droplet size, air temperature, icing time, airfoil angle of attack, and airspeed." The researchers drew no conclusion from the data about the need of interceptor aircraft for icing protection systems. Rather, the purpose of the investigation was to provide the Air Force and manufacturers with a sufficient body of data that would enable them to reach a decision on the matter.[25]

Addressing a conference in Ottawa in June 1955, branch chief von Glahn proclaimed: "Aircraft are now capable of flying in icing clouds without difficulty . . . because research by the NACA and others has provided the engineering basis for ice-protection systems." Ironically, the very success of the NACA's icing research led administrators to question the need to continue the program. If, as von Glahn had indicated, the major problems that remained were engineering in nature, there seemed little need for fundamental research. And the NACA's mission was to do research on the cutting edge of technology; engineering was best left to manufacturers.[26]

Abe Silverstein, now chief of research at Lewis, concluded in the mid-1950s that the NACA's icing research program should be terminated. He believed the icing staff could be better employed elsewhere, as the agency's mission came to emphasize the space program. Although von Glahn and other members of his staff believed that further work remained to be done, they could not quarrel with the basic premise that their

---

[25] Gray and von Glahn, "Aerodynamic Effects Caused by Icing of an Unswept NACA 65A004 Airfoil," NACA TN 4155 (1958).

[26] von Glahn, "The Icing Problem."

fundamental research had succeeded. In any event, Silverstein had made up his mind. At one point, the chief of research offered a heroic solution to terminate work in the IRT when he suggested that a bomb be placed under the tunnel. Although a joke, the comment nonetheless clearly conveyed Silverstein's position on the future of icing research at Lewis.[27]

The IRT's level of activity decreased in 1956 to 555 hours. The next year, usage plummeted to 257 hours. By the end of 1957, all of the NACA's research activity in the IRT had ended, and the icing section's talented staff were transferred to other duties within the renamed (in 1958) National Aeronautics and Space Administration (NASA). Icing research had come to an end—or so it seemed.[28]

---

[27] von Glahn interview.

[28] C. S. Moore, "Major Test Facilities, 1955–1962," n.d., History Office, GRC.

Chapter 4
Industry in the IRT

The IRT, despite NASA's decision to terminate research programs that used the facility, remained attractive to industry as the largest refrigerated icing tunnel in the world. In January and February 1958, Convair tested the wing and tail of its new 880 jet transport in the IRT. Later in the year, Grumman came to the facility to examine ways to protect the large radome of its W2F (Hawkeye) Airborne Early Warning aircraft against icing. Total tunnel usage, however, amounted only to 150 hours in 1958. Activity increased slightly in 1959 with the tunnel running for 200 hours, most of this time being spent on tests of slotted wings for the Douglas DC-8. There were no tests at all in the IRT for the first eleven months of 1960, and only 45 hours of running time in December. As industry seemed finished with the tunnel, NASA placed the IRT in standby status and considered deactivating the facility. Indeed, the IRT might well have been torn down had it not been for the interest of a young engineer from the Vertol Division of the Boeing Company.[1]

In March 1960, the helicopter division of Boeing asked for a volunteer to test the engine inlet of the CH-47 Chinook in the natural icing facility atop Mount Washington. Although he had no experience with icing, Andrew A. Peterson of Vertol's Internal Aerodynamics Group decided that the assignment might be an interesting adventure. After two weeks of tests on the windy 6,250-foot mountain, it was clear that more work was necessary on a system to redistribute engine bleed air to prevent icing. As the Air Force was in charge of the Chinook project, he sought advice from the technical staff at Wright-Patterson Air Force Base. The Air Force recommended that he use the IRT for further tests. Peterson, who had never heard of the facility, was given Uwe von Glahn's name as a contact at Lewis.[2]

---

[1] "Models and Run Days for Major Facilities: Icing Research Tunnel, 1950–1986," n.d.; Ronald J. Blaha, "Completed Schedules of NASA-Lewis Wind Tunnels, Facilities and Aircraft, 1944 to 1990," February 1991; both documents are in the History Office, GRC.

[2] Interview with Andrew A. Peterson by William M. Leary, 22 October 2000.

Figure 4–1. B-24 with Westinghouse 24C-2 turbojet engine mounted under wing, 1948. (NACA C-20866)

Von Glahn, Peterson discovered, was not enthusiastic about the prospect of reopening the tunnel. The IRT, he said, would soon be torn down. Peterson, however, managed to convince him that helicopter icing needs were unique. Flight tests were prohibitively expensive and dangerous. A good deal of valuable work could be accomplished in the icing tunnel. Von Glahn told Peterson that he would take the matter up with his superiors.

NASA finally agreed to allow Peterson to run tests in the IRT. The Boeing researcher then obtained Air Force sponsorship, which meant that no extra money would be required to use the government facility. Peterson went to Cleveland at the beginning of June 1961. When NASA's technicians turned on the tunnel, all the circuit breakers blew. It took three days to check the electrical circuits and get the tunnel running.

Tests of the engine inlet began on 5 June and lasted until 23 June 1966. As a result of this work, Boeing was able to qualify the Chinook's bleed-air anti-icing system. Pleased with the usefulness of the facility, Peterson returned in August for two months of work that resulted in certification by the FAA of the electro-thermal and bleed-air

Figure 4–2. Westinghouse 24C-2 turbojet engine under wing of B-24, 1949. (NACA C-23705)

anti-icing systems of the V-107, a turbine-powered machine that was destined for commercial service with New York Airways.[3]

Peterson used the IRT frequently over the next four years. When he wanted to arrange tests, he would call Lewis and ask for a meeting to review the proposed test plan. He then would draft a letter for a government agency to send to NASA for sponsorship. After contacting Vernon Gray at Lewis, he would set up a schedule for the tests in the tunnel. Initially, NASA conducted the experiments. After the first few tests, however, Boeing's staff ran the tests, while Lewis technical personnel operated the IRT.

Between 1962 and 1964, Peterson conducted tests to qualify the electro-thermal anti-icing system for the engine inlet of the CH-46 (Sea Knight) and CH-113, both derivatives of the V-107. While he was by far the leading user of the facility, he was not

---

[3] Peterson, "Component Icing Tests Conducted at the NASA Lewis (Glenn) Research Center, Icing Research Tunnel," 2 August 2000. Copy provided to the author by Mr. Peterson.

RUN NO.5

C-23984
9-2-49

Figure 4–3. Ice formed on spray nozzles during test, 2 September 1949. (NACA C-23984)

the only one. In 1964, Aerocommander and Learjet tested engine inlets in the tunnel. Indeed, by the mid-1960s, the IRT had become a favorite testing site not only for helicopters, but also for general aviation and large transports. Douglas tested the wing of their DC-9 in the tunnel. Other experiments involved the engine inlet of the DC-10, the radome of the Lockheed C5A, the wing of the Gulfstream G1159, the Beech King Air engine inlet, and the refueling probe of a Sikorsky helicopter.[4]

As part of the FAA certification program for the icing protection system for the slotted wing of the L-1011, Lockheed used the tunnel for a major series of the tests that ran from 5 November through 30 December 1968. A full-scale model of an L-1011 wing section (near the tip) was put through a series of icing conditions to simulate cruise, high-speed hold, descent, and low-speed hold. In all, ninety-five tests were conducted to demonstrate the ability of the system to deliver bleed air to the four outboard leading-edge slats of the L-1011 wing and satisfy FAA certification requirements.[5]

Heavy tunnel usage by industry continued into the 1970s, with Piper, GE, Douglas, Sikorsky, Deutsche Airbus, LTV, Boeing, Learjet, Teledyne, and other companies testing their products under icing conditions. By 1978, companies were competing for tunnel time. That year saw the IRT in operation for 750 hours during 124 days.[6]

It was clear by the late 1970s that icing remained a major concern for aviation. When the Society of Automotive Engineers (SAE) met in Dallas in October 1972, the individuals in the organization who focused on aeronautics decided to set up a panel to investigate ice accretion prediction methods and to identify deficiencies in the current technology. Chaired by Vinod K. Rajpaul of Boeing, the panel reported in May 1978 that little significant work on icing had been done since the NACA research of the 1950s. It concluded that there was an urgent need for a new icing research program to deal with aircraft technology that had changed significantly over the past 20 years. Commercial transport manufacturers were especially interested in developing more fuel-efficient systems in light of rising fuel prices due to the actions of the oil-producing nations (OPEC).[7]

In addition to commercial aviation, the U.S. Army was growing concerned about icing problems. In the late 1960s, as the war in Vietnam wound down, Army aviation had begun to focus on operational difficulties in Europe. One of the major problems was the inability of helicopters and some fixed-wing aircraft to operate in icy conditions. In

---

[4] Blaha, "Completed Schedules."

[5] Lockheed-California Company, "Full Scale Wing Slat Model Anti-Icing Test Program Conducted in NASA 6' x 9' Icing Tunnel," n.d. [1970]; copy in the History Office, GRC.

[6] Blaha, "Completed Schedules;" "Models and Run Days."

[7] Rajpaul, "Aircraft Icing Research Panel Report to SAE AC-9 Committee," May 1978; copy in NASA Conference Publication 2086 (1978), pp. 129–31.

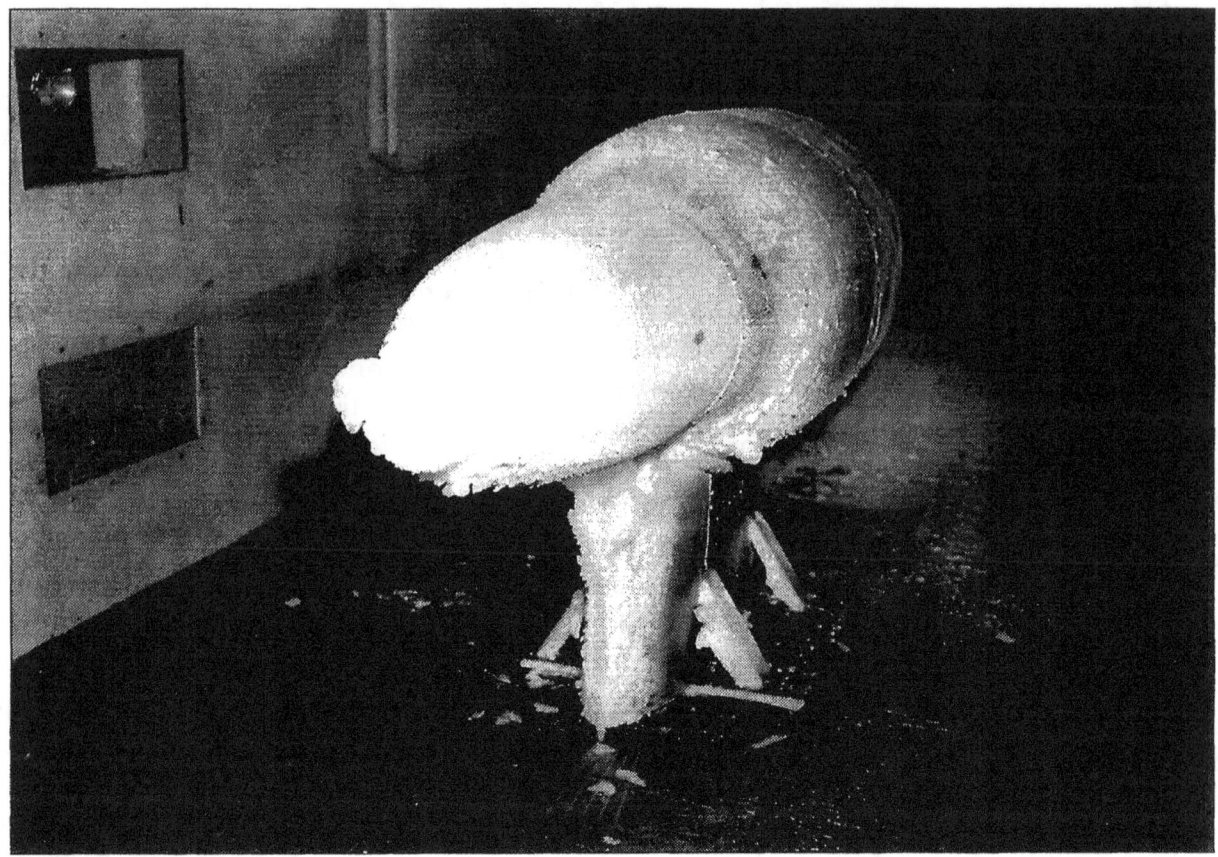

Figure 4–4. Unprotected propeller spinning after icing run, 1948. (NACA C-34635)

1970, Richard I. Adams, as lead engineer for environmental technology, became responsible for aircraft icing research and development programs at the U.S. Army's Applied Technology Laboratory in Fort Eustis, Virginia.[8]

Adams launched an R & D program to establish meteorological criteria for helicopter ice-protection systems. The major technological shortcoming, he early determined, was rotor blade protection. A UH-1H (Huey) was modified to act as a test bed, incorporating experimental ice-protection systems. Initially, the redesigned UH-1N flight-tested a cyclic electro-thermal de-icing concept that the Army had developed. Results proved disappointing, primarily due to excessive weight requirements for the system. "There is a pressing need," he concluded, "to develop advanced concepts for rotor blade ice protection to provide military and civil operators a cost-effective, light-weight means of protecting rotor blades."[9]

---

[8] Adams to Leary, 9 November 2000.

[9] Adams, "Overview of Helicopter Ice Protection System Development," NASA Conference Report 2086, pp. 39–65; Adams to Leary, 19 June 2001.

Figure 4–5. Researcher Dean Bowden inspects pneumatic de-icer on NACA 0011 airfoil model in IRT, 1948. (NACA C-35792)

Adams's concern with the need for helicopter ice-protection systems eventually led to a meeting of government agencies at Fort Eustis on 9 December 1977. Adams informed representatives from NASA and the FAA of the Army's research and development work, which he hoped would provide the basis for establishing a broader program of icing research.[10]

Milton A. Beheim, a division chief at Lewis who attended this meeting, embraced the idea of NASA's participation in the icing program. He gave his enthusiastic endorsement to Bernard Lubarsky, deputy director at Lewis, for reopening the icing section. Lubarsky supported the recommendation and obtained the approval of Headquarters.[11]

To launch NASA's reinvigorated icing effort, Beheim decided to hold a workshop at Lewis that would bring together the various agencies and individuals that would later

---

[10] Adams to Leary, 12 November 2000.

[11] Interview with John J. Reinmann by William M. Leary, 23 September 2000.

Figure 4–6. Researcher Thomas Gelder with spray system for dyed water impingement studies in IRT, 9 September 1949. (NACA C-49-24015)

constitute the "in flight icing community." Sponsored jointly with the FAA, and with planning assistance from the Department of Defense, the workshop met on 19–21 July 1978. Participation exceeded all expectations. In all, 113 individuals from industry, the military, and government agencies attended, including representatives from the Netherlands, United Kingdom, France, Sweden, and Norway.

Had this meeting been held a few years earlier, Lubarsky observed in his welcoming remarks, "I believe we could have all sat around a card table." But this had changed. Requests for testing in the IRT had reached a point where they were difficult to satisfy. This renewed interest in icing, Lubarsky said, was being driven by new requirements and by the opportunities offered by new technologies. Commercial transport manufacturers wanted more efficient equipment. General aviation interests wanted to certify their aircraft to fly into forecast icing conditions. There was a similar need for helicopters, stemming from military requirements to operate throughout Europe during the winter, and from commercial needs to service oil rigs in the North Sea and off the Alaskan coast.[12]

John H. Enders, from the Safety Office in NASA Headquarters, then made the introduction. The intent of the workshop, he noted, was to share information from a wide range of specialist groups, most of whom normally have little or no regular contact with one another. The participants in the workshop then toured the IRT and the Altitude Wind Tunnel. Enders told them that NASA was considering a proposal to modify the AWT into a large subsonic propulsion and icing test facility.[13]

Conference participants were then assigned to committees on meteorological research, icing forecasting, systems development, civil operations, military operations, and icing research and facilities. Over the next day and a half, the committees considered all aspects of their area of responsibility, and then they rendered a final report of recommendations.

Beheim delivered the executive summary of the workshop. He began by reviewing the past, and then he looked to future developments in icing research. Between 1940 and 1955, he said, industry and government laboratories, especially the NACA, had made a significant effort to define the natural icing environment, determine its effect on aircraft components, and design technologies for ice protection. The results of this work established the basis for certification requirements for large transports. This effort had ended in the late 1950s, as transports shifted to jet power and had ample supplies of engine-bled air to cope with icing during ascents and descents.

Although it is clear that the old NACA research was held in high regard, Beheim continued, the lack of serious efforts to advance the state of the art since 1957 had resulted in

---

[12] Lubarsky, "Welcoming Remarks," NASA Conference Publication 2086, pp. iii–iv.

[13] Enders, "Introduction," ibid., pp. vii–ix.

Figure 4–7. Icing researchers Donald Mulholland (left) and Vern Rollin (right) stand between a Canadian researcher during a visit to Canada's National Research Council, December 1949. (NACA C-24930)

serious deficiencies in design tools and regulatory criteria for aircraft that differed significantly from those of the 1950s, especially rotorcraft and the fixed-wing airplanes used by general aviation. Workshop participants had "sharply underscored an essential NASA role to accomplish the following objectives: update the applied technology to the current state of the art; develop and validate advanced analysis methods, test facilities, and icing protection concepts; develop improved and larger testing facilities; assist in the difficult process of standardization and regulatory functions; provide a focus to the presently disjointed efforts within U.S. organizations and foreign countries; and assist in disseminating the research results through normal NASA distribution channels and conferences."

The workshop, Beheim concluded, had reinforced NASA's growing awareness of the demands for renewed icing research. In view of this "obvious need," NASA Headquarters had approved the reestablishment of an icing research section at Lewis. NASA, he announced, would commission a series of study contracts with industry to define the program for a larger research effort in the future. Also, plans to modernize the IRT had been initiated. Finally, feasibility studies were underway for the design of a much larger icing tunnel with higher speeds and variable altitude capability; although, the eventual outcome of these studies to convert the AWT "will be highly dependent on budgetary restraints."[14]

To manage the IRT, Beheim selected Harold Zager. Already a section head in Beheim's Wind Tunnel and Flight Division, Zager's responsibilities included the 8' x 6', 9' x 15', and 10' x 10' wind tunnels. But he had never before seen an icing tunnel. His first look at the facility was not encouraging. The IRT had been used rarely and was not being maintained. Fred Murray, an experienced mechanic, had been looking after the tunnel, but he had lacked the authority to get much done. The ceiling had rotted and had fallen down on the air exchange doors, which could not be moved. Zager told Beheim that he was not prepared to take responsibility for the IRT until repairs had been made so that it could be safely operated. It was "a good time" to make such a request, Zager recalled, and Beheim agreed to the necessary expenditures.[15]

Beheim, with the assistance of branch chief Roger Luidens, also recruited the first four researchers to staff the tunnel. To form the core group of researchers, Beheim used his power of "persuasion" on William A. Olsen, Jr., a talented experimentalist; Bernard J. Blaha, an accomplished wind tunnel research engineer; Peggy Evanich, an experienced aeronautical engineer who had worked for Lockheed before joining NASA; and Robert J. Shaw. Shaw's experience was more-or-less typical of what happened. Shaw recalls that Beheim called him into the office shortly before the workshop concluded and asked him to become part of the icing research group that was being formed. There was no money at present, Beheim allowed, and the decision had been made to start out as a section rather than at the higher level of branch. (The section would become a branch in 1987.) Shaw, who was working on V-STOL inlets, was not exactly excited at the prospect of a transfer. Icing was not in the mainstream at Lewis. During the 1970s, however, one did not turn down "requests" by management. "You did what you were told," Shaw commented. As it turned out, the transfer to icing research became "one of the best breaks of my career." It was "a fascinating experience," he recalled in 2001, and "the most fabulous ten years of my thirty-one-year career with NASA."[16]

Beheim still had to find someone to head the section. Advertisements for the position produced a number of applicants. In the end, he opted for John J. Reinmann. This would prove a happy choice. After graduating from Case Institute with a B.S. (1954) and M.S.

---

[14] Beheim, "Executive Summary of Aircraft Icing Specialists Workshop," ibid., pp. 1–12.

[15] Leary interview with Zager, 27 June 2001.

[16] Leary interview with Robert J. Shaw, 26 June 2001.

Figure 4–8. Carrier components and condensors of the refrigeration plant, 5 February 1950. (NACA C-25219)

(1957) in mechanical engineering, Reinmann went on to receive an M.S. in physics from John Carroll University. He then took a position with the New Devices Laboratory of TRW, Inc., attaining the position of project engineer while working on the development of Rankine power conversion systems for space satellites. Reinmann joined NASA in 1961 and spent "seventeen glorious years" doing research into plasma physics and thermonuclear fusion for space propulsion, with fifteen of those years serving as head of the High Temperature Plasma Section. In 1978, however, the Lewis laboratory experienced the worst reduction in force in the history of the facility. Bruce T. Lundin, director of Lewis, announced that he was not interested in any research that did not pay off in five years. As a result, Reinmann and other researchers in his section were soon out of a job—and without "bumping" rights.

NASA offered Reinmann only one option: to join a new computational fluid mechanics training program. Although this would mean a reduction in salary, the need to support a large family left Reinmann with no choice but to accept. On a more positive note, he looked forward to acquiring expertise in the field of computational fluid

Figure 4–9. Refrigeration plant for the IRT and AWT, 1 March 1951. (NACA C-27262)

mechanics. Just as he completed the nine-month program, Reinmann saw the announcement for a section head of icing research. "After looking into it a bit further," he recalled, "I became excited about it. First, it was a field in which I felt very comfortable: thermodynamics, fluid mechanics, and heat transfer. It was like coming home. Second, it had the Icing Research Tunnel, which meant I could bring analysis and experiments along together. Third, I felt that with my previous analytical and experimental research experience, and with my newly acquired capabilities in computational fluid mechanics, I was eminently qualified." Finally, the position would mean a return to his previous pay grade as section head. "I had nine children to feed, clothe, and send to college," he pointed out. "So, believe me, I didn't take this lightly."[17]

Part of the Low-Speed Aerodynamics Branch (Luidens) of the Wind Tunnel and Flight Division (Beheim), the new section's first task was to write the specifications for

---

[17] Reinmann interview; Reinmann to Leary, 1 April 2001.

Figure 4–10. Vernon Gray inspects the airfoil model during cyclic gas de-icing experiments, 26 September 1951. (NACA C-28406)

the three study contracts that Beheim wanted to commission. Larry P. Koegeboehn of the Douglas Aircraft Company was selected to write the report on the icing research requirements of commercial aviation. B. K. Breeze and G. M. Clark of Rockwell International would survey the light transport and general aviation icing research requirement. Finally, Peterson, Leo Dadone, and David Bevan of Boeing Vertol would handle rotorcraft icing needs.[18]

Breeze and Clark were the first to finish. General aviation and light transport aircraft, they pointed out in their report, had a higher exposure to icing conditions than large transports. Furthermore, they paid greater penalties for ice-protection systems due to their smaller dimensions. A major effort was needed to resolve the icing problem with the aid of new technologies. "It becomes increasingly apparent that in order to obtain the wind tunnel database required to solve general aviation aircraft problems," the authors continued, "the NASA icing wind tunnels will have to be used to a greater degree, and improvements will have to be made to expand the applicability of these tunnels, increase the accuracy of test measurements, and reduce the turnaround time between tests."

Breeze and Clark recommended that the IRT be improved with modern recording and data-processing equipment, and that the AWT be converted into an icing tunnel. The report went on to list thirty-three areas of desired research. Although there was no industry consensus on research priorities, there was general agreement that a training film for flight in icing conditions would be beneficial. Finally, there was a consensus that NASA Lewis should be "the center of aircraft icing expertise for basic research and consultation, and should act as a clearing house for exchange of information for industry involvement."[19]

Koegeboehn next weighed in with the views of the commercial aircraft industry. He agreed with the previous report's recommendation that the IRT be modernized with an automated control system and that the AWT be converted into an icing tunnel. Noting that rising fuel prices in recent years had encouraged aircraft designers to seek ice-protection systems that would save weight and fuel, he proposed, in order of priority, four research goals that should be reached within four years. First, NASA should develop and test an electro-impulse de-icing system that would save fuel and be low in cost. Second, super-critical airfoils should undergo ice accretion tests in the IRT, with measurement of drag and lift coefficients. Third, NASA should test various devices to measure cloud properties and recommend acceptable types to the military and FAA. Finally, a new computer program for ice-protection analysis should be developed.

---

[18] Reinmann interview.

[19] Breeze and Clark, "Light Transport and General Aviation Aircraft Icing Research Requirements," NACA Contractor Report (CR) 165290 (March 1981).

Figure 4–11. Bank of manometer tubes used to record pressure on a model in the IRT test section. (NACA C-161503-5)

Long-range research goals, Koegeboehn went on, should include a program to evaluate new materials with ice-phobic properties, develop and test microwave ice-protection systems, and update or develop a new handbook of icing technology.[20]

The recommendations of the rotorcraft industry contained few surprises. The current emphasis being placed on helicopter icing by the Department of Defense and the FAA, it noted, stemmed from the increased need for all-weather clearance for helicopters, the desire to increase the overall utility of rotorcraft without weather restrictions, and the need to safely operate helicopters over regions with a wide range of weather conditions, including those without good forecasting abilities. Despite the need, helicopter icing research had progressed very slowly over the past twenty years. As tests in the IRT offered the most comprehensive methods of determining the performance of non-

---

[20] Koegeboehn, "Commercial Aviation Icing Research Requirements," NASA CR 165336 (April 1981).

rotating ice-protection systems under varied conditions, the report recommended that the "upgrading of existing major icing test facilities must continue as rapidly as funds permit." The IRT could be modernized to improve the range of liquid water content and droplet size to meet civil and military helicopter requirements. Also, the AWT should be reactivated as an icing tunnel.

The report emphasized that the helicopter rotor was the "key area" requiring a major effort. Although the IRT was not large enough to accommodate a full-scale rotating main rotor, work was already underway to test an OH-58 tail rotor rig in the tunnel. These tests offered the opportunity to determine the feasibility of using scale-model rotors for testing.[21]

Reinmann was pleased with the three studies, which generated tremendous interest both at Lewis and at NASA Headquarters. They not only enabled NASA to put together a responsive program with well-defined objectives, but they also made everyone aware that NASA "was back in the icing business." The main problem with implementing the program, however, was lack of funds.[22]

For the first three years, Reinmann had to work with a limited annual budget of only $500,000. But the situation improved in 1981, thanks largely to the interest of Allen R. ("Dick") Tobiason. Trained as an engineering test pilot and a retired Master Army Aviator, Tobiason became the manager of aviation safety technology at NASA Headquarters in 1980. "I found many low-funded programs," he recalled, "that did not appear to have the potential for improving safety in the short term, nor were there prospects to increase total funding." He decided to cancel several of the smaller programs and use the money to increase funding for programs which had more immediate prospects for improving safety and which used unique NASA facilities. One of the major recipients of Tobiason's largess was icing research.[23]

Tobiason not only quadrupled icing funding, but he also arranged for the transfer to Lewis of a Twin Otter research aircraft. "I wanted," he noted, "some test airplanes involved to prove theory, wind tunnel results, and analytical models." The Langley laboratory, where the Twin Otter was based, was reluctant to part with the airplane. Tobiason, however, had the necessary "clout" to override Langley's objections. The transfer included an agreement between Lewis and Langley to share data.

Finally, Tobiason wanted closer relations between Lewis and the FAA, which was in the process of revising icing standards for certification. Largely through his

---

[21] Peterson, Dadone, and Bevan, "Rotorcraft Aviation Icing Research Requirements," NASA CR 165344 (May 1981).

[22] Reinmann interview.

[23] *Lewis News*, 10 April 1981; Tobiason to Leary, 10 July 2001.

efforts, a cooperative agreement was signed, resulting in the transfer of funds from the FAA to Lewis and the assignment of an FAA staff representative to the Cleveland laboratory.

Tobiason's efforts were certainly appreciated by Reinmann. One of the few people at Headquarters to "put his money where his mouth was," Tobiason represented "a breath of fresh air" for icing researchers. And his emphasis on safety issues proved fortuitous, as a dramatic accident would soon focus national attention on icing problems.[24]

---

[24] Reinmann to Leary, 11 July 2001.

Chapter 5

Back in Business

On the morning of 13 January 1982, Washington, DC, was in the midst of a major snowstorm that would dump 6 inches on the nation's capital. Schools and businesses closed; Congress recessed early; and National Airport was shut down. By noon, as the steady snowfall gave way to scattered showers, the airport reopened. Just before 1:40 P.M., however, it closed again to allow plows to clear the instrument runway (18/36). It was scheduled to resume operations at 2:30 P.M.

As the plows did their work, passengers boarded Air Florida flight 90, a Boeing 737-222B, that was bound for sunny Ft. Lauderdale. Captain Larry Wheaton asked for de-icing as the seventy-four passengers, no doubt happy at the prospect of escaping the northern winter, found their seats. The de-icing was started, but then it stopped after Wheaton was informed that there would be a further delay, and that he would be eleventh in line for departure. At 3:00 P.M., the airplane was de-iced for 10 minutes. Thirteen minutes after that, Wheaton was cleared to push back from the gate. The tug, however, its tires spinning in the snow, could not move the Boeing 737. Wheaton suggested that he use the aircraft's thrust reversers to gain some momentum. Although engaged for about 1.5 minutes, the reversers succeeded only in throwing up slush and snow. The aircraft remained immobile. Finally, a second tug, equipped with chains on its tires, managed to push back the 737.

Palm 90, the aircraft's call sign, then taxied out behind a New York Air DC-9 and became number seventeen in line for takeoff. A light snow continued to fall as Palm 90 waited to depart on runway 36. Finally, at 3:59 P.M., the control tower gave Wheaton clearance to take the runway. Visibility at this time was down to 0.25 of a mile.

Seconds later, the tower gave Palm 90 permission to depart. As the aircraft reached rotation speed, the nose of the 737 pitched up sharply. "Forward, forward, easy," Wheaton told copilot Roger Pettit. But the aircraft did not respond. "Come on. Forward, forward," Wheaton exclaimed. "Larry, we're going down," copilot Pettit responded. "I know it," were Wheaton's last words, just before Palm 90 slammed into the 14th Street Bridge.

The Boeing 737 crushed four automobiles on the busy bridge, killing five people. The airplane then hit the ice-covered Potomac River, sinking up to its tail. Seventy passengers and four crewmembers died in the crash. Only one flight attendant and four passengers were pulled from the wreckage in a dramatic rescue that received full television coverage.

The National Transportation Safety Board listed a number of causes for the accident, but assigned major responsibility to icing. The NTSB report faulted the flight crew for failing to turn on engine anti-ice during ground operations and takeoff, their decision to take off with snow and ice on the airfoil surfaces of the 737, and the prolonged ground delay between de-icing and receipt of takeoff clearance, during which the aircraft was exposed to constant precipitation.[1]

This tragic and dramatic accident, which received the widest possible media coverage, thrust the dangers of icing to the forefront of national attention. It certainly had a major impact on NASA's icing research program. Reinmann, who had been hard pressed to show icing as an important safety issue, now had a vivid example of the need for additional research for safety purposes. It became easier for him to argue for funds for the icing section. Eventually, the impact of the Air Florida accident eased the way for Reinmann to obtain approval for a major renovation of the IRT.[2]

Even before the Air Florida accident, Reinmann had begun research programs that responded to the needs of industry for better methods of ice protection. For a time, it appeared that the answer to the vexing problem of helicopter rotor blade de-icing might well lie in an improved version of the Goodrich de-icer boot. One of the oldest ice-protection methods, the pneumatic boot was attractive as a low-weight, low-cost, and low-power method of ice removal. It had been studied by Lockheed Aircraft under an Army contract during the 1970s. The concept had been rejected, however, because the neoprene rubber of the boot could not withstand heavy rain and could be damaged or completely torn off by centrifugal force.[3]

Goodrich investigated a variety of materials to improve the durability of the boots. The company finally became convinced that polyurethane (trade name, Estane) would solve the problems that had been associated with neoprene. In 1979, Goodrich and NASA conducted a joint program to test the polyurethane boot in the IRT. Placing a 6-foot segment of a full-scale Bell UH-1H (Huey) rotor blade in the

---

[1] National Transportation Safety Board Report NTSB-AAR-82-8, 10 August 1981.

[2] Reinmann interview.

[3] Robert J. Shaw, John J. Reinmann, and Thomas L. Miller, "NASA's Rotorcraft Icing Research Program," preprint for the NASA/Army Rotorcraft Technology Conference, NASA Ames Research Center, 17–19 March 1987; copy in the History Office, GRC.

Figure 5–1. Overhead view of the IRT and AWT, 1971. (NASA C-71-3281)

tunnel, researchers Blaha and Evanich tested three boot configurations. They found that the most successful configuration had two spanwise tubes surrounding the leading edge and chordwise tubes aft of the leading edge on both suction and pressure surfaces. Their proof-of-concept tests were sufficiently encouraging to lead to a flight test program.[4]

Sponsored jointly by Goodrich, NASA, the Army, and Bell Helicopter, the flight program began with hover tests at the Canadian National Research Council's Icing Spray Rig. Located at the northwest corner of Uplands International Airport, near Ottawa, the rig consisted of 161 steam-atomizing water nozzles that were mounted on a steel frame. The nozzles produced a cloud that was 75 feet wide by 15 feet deep. A wind velocity of 10 miles per hour was required to move the cloud away from the array. It would cover the main and tail rotors of a hovering helicopter. While useful as a "first look" technique for the proving of a new de-icing system, the spray rig had a number of limitations. For

---

[4] Blaha and Evanich, "Pneumatic Boot for Helicopter Rotor Deicing," NASA CP-2170 (1980).

Figure 5–2. Rockwell factory representative inspects Hellfire missiles during IRT test, 1978. (NASA C-78-1039)

example, it depended upon ambient temperature to produce icing. Also, tests could only be conducted on a hovering rotorcraft. Nonetheless, the performance of the boot-equipped UH-1H encouraged further flight testing.[5]

Forward flight tests of the Huey were made behind the Army's Helicopter Icing Spray System (HISS) tanker. The spray aircraft was equipped with an internally mounted 1,800-gallon water tank and an external spray boom, suspended 19 feet beneath the helicopter. Tests were usually conducted at 90 knots, with the UH-1H positioned 250 feet behind the HISS helicopter. At this distance, the spray cloud was

---

[5] On the Icing Spray Rig, see T. R. Ringer, "Icing Test Facilities in Canada," Advisory Group for Aerospace Research & Development [AGARD] Advisory Report 127 (1978).

35 to 40 feet wide. Although it was difficult to maintain the correct liquid water content and droplet size with the HISS system, the tests still provided significant information about the inflight performance of the de-icing system. Perhaps the most important result of the HISS tests related to the durability of the Estane boots. After 6 hours of artificial rain testing behind the tanker, there was no detectable erosion of the Estane[6].

Erosion testing continued at Fort Rucker, Alabama, where the Estane boots withstood 10 hours of sand erosion. However, 10 hours of flying in natural rain required four separate blade repairs due to the pitting and chunking of the Estane. By 1987, the conclusion had been reached that the Goodrich boot was a feasible rotor-protection system, but that additional field tests were required to gain more experience with the erosion problem.[7]

Far more promising than the Goodrich boot as an innovative ice-protection system for both helicopters and fixed-wing aircraft was the Electro-Impulse De-Icing (EIDI) system that occupied a good deal of NASA's attention during the 1980s. The use of electro-magnetic energy as an impulse force to remove ice had been first suggested in 1937 by Rudolf Goldschmidt, a German national who was living in London. Although Goldschmidt had obtained a patent for his concept, he apparently had never attempted to build a device to test his ideas. The concept had emerged again in the mid-1960s in the Soviet Union. I. I. Levin of the Ministry of Power and Electricity had published a paper on the possibility of employing electro-impulse force to remove frozen and sticky material from bunkers, transformer boxes, and other surfaces.[8]

The Soviets had installed the system on an Ilyushin IL-18 transport in the 1970s, then sent salesmen to Western nations to market the concept. They had shown a film of a wing with heavy icing. In an instant, the ice had disappeared. Although it had been impossible to tell whether the wing was attached to an airplane in flight or had been photographed in a wind tunnel, viewers had nonetheless been impressed. The Soviets had refused to provide any further details, had referred interested parties to Levin's article, and had asked for $500,000 to develop the system for any aircraft.

The Russians had not gotten their $500,000, but their visit had sparked interest in the system in Western Europe and in the United States. The French had attempted to put an electro-impulse system on an Alpha Jet, but it had not worked well. The British

[6] "NASA's Rotorcraft Icing Research Program." The HISS system is described in the FAA's Aircraft Icing Handbook, vol. 2, ch. IV.
[7] "NASA's Rotorcraft Icing Research Program."
[8] For the EIDI story, see G. M. Zumwalt, R. L. Schrag, W. E. Bernhart, and R. A. Friedberg, "Electro-Impulse De-Icing: Testing, Analysis and Design," NACA CR 4175 (September 1988); and Zumwalt to Leary, 30 August 2000.

Figure 5–3. Technician Fred Murray with traverse rig in IRT, 1979. (NASA C-79-1686)

Aircraft Corporation had tried the system in a wind tunnel, but they had found that the force required to remove ice also destroyed the wing structure. In the United States, Bell Helicopter and Lockheed had looked into the concept. Lockheed's experiments had appeared especially promising, but the company's decision to leave the commercial transport business had ended work on the project.

Aware of Lockheed's apparent success, NASA invited McDonnell Douglas to test EIDI in the IRT in 1981. In November, a semi-trailer arrived in Cleveland from California with fourteen technicians and engineers, impressive capacitor banks, multiple recording devices, and a DC-10 wing section with removable leading edges and variable-location pads for the copper coils that would carry the electro-magnetic energy. McDonnell Douglas tested the EIDI-equipped wing section in the IRT for six weeks. Sometimes the system worked, and sometimes it did not. No one knew why. With the results deemed negative, the McDonnell Douglas people packed up and went back to California. The data from the experiments were never reduced, so NASA learned nothing from the tests. In any event, most individuals connected with the tests came away with the conviction that EIDI was a fluke.

Reinmann, on the other hand, continued to see promise in EIDI. He sought to interest a Lewis colleague, who was an expert in electro-magnetic theory, in undertaking further research into the system, but he was turned down. Reinmann next approached a professor at MIT who had published extensively on the interaction of electro-magnetic waves with various media. After talking to several of his graduate students, the professor said that no one was interested in pursuing research in this area. Work on the system seemed at a dead end.[9]

Shortly after these disappointing results, however, EIDI again came to NASA's attention—this time during a meeting of the General Aviation Advisory Committee at Lewis. A representative of a small aircraft company complained about the limitations of pneumatic boots and asked if NASA could come up with a better de-icing system for piston-engine aircraft. Icing researcher Harold W. Schmidt, who was an avid private pilot, said that EIDI might be the answer. But, he went on, NASA was reluctant to work with the large transport manufacturers on developing the system because of the recent experience with McDonnell Douglas. Representatives from Cessna and Beech, aircraft companies in Wichita, Kansas, then suggested that a study might be done through a local university, with the results made public.

Following this meeting, Schmidt called Professor Glen W. Zumwalt at Wichita State University (WSU) and told him that NASA might be prepared to work with Cessna and Beech if they were partners in a quick feasibility study of EIDI that was proposed by a

---

[9] Reinmann to Leary, 22 April 2001.

Figure 5–4. Inspecting IRT fan blades, 1979. (NASA C-79-324)

university.[10] Zumwalt was at first hesitant to become involved. He reminded Schmidt that he was an aerodynamicist who was familiar with wind tunnels, but that he had no background in electro-dynamics or structural dynamics. Schmidt, however, would not be fended off. In the end, Zumwalt agreed to explore the idea with the two individuals who had brought up the subject at the Advisory Committee meeting—David Ellis of Cessna

---

[10] David Ellis, who was chief of advanced design at Cessna, has a somewhat different recollection of these events. He recalls that his first knowledge of the EIDI project was the result of a telephone call from Reinmann about this "Russian system." Reinmann told Ellis that NASA had tried to interest Boeing, Douglas, and Lockheed in a joint project with NASA to develop EIDI, but they were not interested. Reinmann wondered if a general aviation manufacturer might wish to join such a project. Ellis told Reinmann that Cessna would indeed be interested. Ellis then contacted Zumwalt, who set up the WSU team. Leary telephone interview with Ellis, 1 September 2000. Zumwalt, however, responded to Ellis's recollection in caps: "I AM QUITE SURE THAT I HEARD FIRST FROM SCHMIDT, THEN CONTACTED DAVE ELLIS." Zumwalt to Leary, 23 May 2001.

and Harold Reisen of Beech. It did not take long to put together a proposal for $70,000 to study and design an EIDI system that would be installed on a Cessna wing and a Beech wing of similar size with different leading edge stiffness and shape. The wings would be tested in the IRT within six months. This proposal was submitted to NASA in mid-February 1982 and was promptly approved. The contract specified that testing would begin no later than October 1982.

Zumwalt quickly put together his interdisciplinary EIDI team. For the electro-dynamics, he recruited Robert L. Schrag, professor of electrical engineering at Wichita State. Next, for the structural dynamics, he signed on Walter D. Bernhart, WSU professor of aerospace engineering. Zumwalt then made a quick trip to Cleveland to learn about the IRT. While there, he was informed that Simmons-Precision, an aircraft electrical system manufacturer, had shown an interest an EIDI. He contacted the company and invited them to furnish the power supply for the system. After a visit to Wichita, two engineers from Simmons-Precision agreed to join the WSU team.

Zumwalt adopted the basic EIDI system that had first been tested in the 1970s. It consisted of flat-wound coils of copper ribbon wire that were placed just inside the leading edge of the wing's skin, with a small gap between the skin and the coil. The coils were then connected to a high-voltage capacitor bank. Energy was discharged through the coil by a remote signal to a solid-state switch, a "silicon-controlled rectifier" (SCR). The discharge created a rapidly forming and collapsing electro-magnetic field that induced eddy currents in the metal skin. The result was the creation of a repulsive force of several hundred pounds for a fraction of a millisecond. This force shattered, debonded, and expelled the ice—instantaneously. Zumwalt had "a clever graduate student," Robert A. Friedberg, fabricate the copper coils and mount them on the Cessna and Beech wing sections.

On the weekend of 23–24 October 1982, Friedberg drove the models from Wichita to Cleveland in a rental truck, while Zumwalt, Schrag, Bernhart, and representatives from Cessna and Beech flew to the Lewis laboratory for the tests. Monday morning, 25 October, the WSU-led team met with the staff of the icing section. "We were patronized," Zumwalt recalled. "If all of the so highly regarded organizations have failed to make EIDI work," they let Zumwalt know, "we will understand if little Wichita State and two dinky light airplane companies fail." Only Schmidt and Reinmann remained optimistic.

Tests of the Cessna 206 wing section began on Monday evening. EIDI performed flawlessly. The wing was successfully de-iced through a range of speeds, temperatures, angles of attack, and droplet sizes. Schmidt was ecstatic. He invited Reinmann to view the next night's tests. Again, on Tuesday evening, EIDI worked its magic. Reinmann, Zumwalt remembered, "was jumping up and down saying 'Just like those Russian movies . . . Just like those Russian movies!'"

Figure 5–5. Researcher Peggy Evanich holds a Cessna wing section in the IRT, 1980. (NASA C-80-4021)

A second week of testing involved the Beech Bonanza wing section. Again, there were only positive results. Two weeks of IRT testing had shown that EIDI could de-ice two different general aviation wings with a low-energy expenditure of 800 joules per foot of span during the de-icing cycle. The key to designing EIDI, Zumwalt later recalled, proved to be "getting the electro-dynamicist and the structural-dynamicist to talk to each other," as there were strict time interval requirements for both phenomena, and they had to interact correctly.[11]

Zumwalt had met with NASA officials in the midst of these successful tests and had suggested a three-year program during which WSU would install and test in the IRT the EIDI system on a variety of fixed- and rotor-wing aircraft. In addition, there would be flight tests of the system on the NASA Twin Otter and on a Cessna 206. With Reinmann's enthusiastic endorsement, NASA speedily approved Zumwalt's proposal. Zumwalt would be project director with Schmidt as project manager. The initial members of the consortium of manufacturers were Beech, Cessna, Gates-Learjet, LearFan, Boeing, McDonnell Douglas, and Simmons-Precision. They later were joined by Rohr, Sikorsky, Bell, and Boeing-Vertol.

The first series of tests in the IRT for the new program took place between 18 and 22 April 1983. There were thirty-one runs made to compare different coil-and-mount designs on a Cessna 206 wing section. In August, a 6-foot section of Twin Otter wing was used for thirty-five icing runs. Some runs were made with continuous icing for up to 21 minutes, with EIDI impulses 3 to 6 minutes apart. De-icing improved after the first impulse sequence in all cases. November brought additional tests of the Twin Otter wing, plus a 38-inch section from a LearFan wing that was made of Kevlar epoxy composite. By this point in time, the energy required to de-ice a general aviation wing had dropped dramatically from 800 joules a foot to 180 joules—comparable to the amount of energy required to power the landing lights.

The next year, 1984, saw a series of four test periods in the IRT during which EIDI was observed on a Learjet wing, Cessna tail section, Boeing 767 slat, Falcon Fanjet engine inlet, and Bell Cobra helicopter blade. In all cases, EIDI successfully removed ice without a problem. Flight tests during the year were equally satisfactory. In January, the NASA Twin Otter, flown by Richard J. Ranaudo and Robert McKnight, made twenty-one flights over Lake Erie while WSU's Friedberg operated the EIDI equipment. EIDI, the pilots reported, worked better than adjacent pneumatic boots. In February and March, an EIDI-equipped Cessna 206 made fifteen flights over Wichita, following a Cessna 404 tanker with spray gear. Again, the EIDI system worked without a hitch.

Results from all the EIDI tests were so promising that Reinmann decided to host a symposium at Lewis in June 1985 to review the conclusions of the three-year

---

[11] Zumwalt to Leary, 23 May 2001.

Figure 5–6. The IRT Control Room, 1973. (NASA C-1987-473)

NASA/WSU/industry consortium. More than 100 individuals from various companies, including all ten companies in the EIDI consortium, and government agencies attended the gathering. Zumwalt and his colleagues delivered a report on the program, displayed an EIDI-equipped Cessna 206, and invited participants to observe tests in the IRT. Throughout the day, the advantages of EIDI were highlighted. The system operated on low energy, caused no aerodynamic penalties, required minimum maintenance, and compared favorably in terms of weight and cost with existing de-icing systems. Cessna announced that it planned to apply in 1986 for FAA certification for the 206 with the EIDI system. "In all," Reinmann concluded, "we believe that EIDI will prove to be a feasible system for application to all aircraft classes."

Hailed as the de-icing system of the future, EIDI never lived up to its promise (at least, to date). A combination of circumstances worked against the system. What might be viewed as an ill omen for the future took place three weeks after the symposium when Zumwalt fell through the ceiling of an underground cave while walking in a national park in California. He broke his back and would never walk again; although, he would continue to teach for seven more years before retiring. Cessna, which had been

Figure 5–7. Computerized IRT Control Room, 1987. (NASA C-1987-969)

the most enthusiastic supporter of EIDI, soon shut down the production lines for their single-engine piston aircraft in the face of prohibitive liability insurance costs. Beech downsized. LearFan went out of business. Schmidt retired from NASA, and the WSU EIDI team scattered. Adoption of the system was impeded by the fact that it could not be retrofitted to existing aircraft and had to be built in during the manufacturing process. Also, the high costs of getting a new system certified by the FAA discouraged manufacturers from opting for EIDI. By the time Zumwalt and his team published their final report in 1988, they sensed that there might well be a hiatus before any new aircraft used the system. "We tried to write the final report," Zumwalt stated, "so that it could provide all that is needed to design and install an EIDI system without any of our old team. I think we did."[12]

The EIDI program was in many ways a model for the interdisciplinary team approach and for the type of NASA-industry-university cooperative effort that would

---

[12] Ibid.

Figure 5–8. Raymond Sotos (kneeling) and David Justavick with Hughes A-64 inlet after icing test, 1980. (NASA C-80-1729)

become the standard for the post-1978 IRT. With limited funding, the icing branch was compelled to seek a variety of partnership arrangements in order to obtain the maximum results from minimum expenditures. "As I look back," Reinmann pointed out, "I now consider our limited support a blessing in disguise. It forced us to develop joint programs with a wide spectrum of industry and with other government agencies (the FAA, Air Force, Army Navy, British RAE, French, and Canadian). It forced us to avoid duplication of equipment, facilities, and expertise. We learned to bring interested parties together and to get them to put their resources on the table. And finally it forced us to go out and find more funding from other sources, by understanding industry's and other government agency's needs, by showing them the advantages of not duplicating efforts." Long before the approach became a Headquarter's mantra, Reinmann emphasized, "we were already engaged in the 'better, faster, cheaper' way of doing applied research."[13]

Facility manager Zager had the responsibility for scheduling the IRT. His first priority was to obtain maximum usage of the tunnel. "My job," he emphasized, "was to keep the tunnel busy." Zager had instructions from deputy director J. M. Klineberg and associate director Seymour C. Himmel "that I was to make the companies happy." Although contractors did not receive priority on scheduling, once their experiments had begun, Zager usually would extend their tunnel time in order to complete the tests. This caused schedules to slip and gave rise to complaints. Zager and Reinmann clashed more than once over scheduling priorities, but the facility manager rarely gave way. On one occasion, Zager recalled, he had to go the laboratory's top management for support—which he received.[14]

Thanks in large part to Zager's determined efforts, IRT usage reached 1,134 hours in 1985, the first time it had exceeded 1,000 hours since reopening. Indeed, as Reinmann observed at the time, the IRT had become "one of the most heavily scheduled tunnels within NASA." The facility, however, was badly in need of renovation. Furthermore, it had become clear that plans to convert the Altitude Wind Tunnel into an icing research facility were going to fall through.

When NASA resumed its program of icing research in 1978, officials at Lewis, as well as potential users, had agreed that conversion of the unused AWT would be a fine

---

[13] Reinmann to Leary, 22 April 2001.

[14] Zager interview. Reinmann has a somewhat different recollection. "I had a lot of support from Beheim," he commented, "but I don't think I ever had to go to him to counter Zager's scheduling. I emphasized teamwork, and we met regularly with Zager and the IRT operators to constantly optimize the scheduling process. In any highly scheduled facility there was bound to be some squabbling about getting time in the facility." Reinmann to Leary, 17 July 2001.

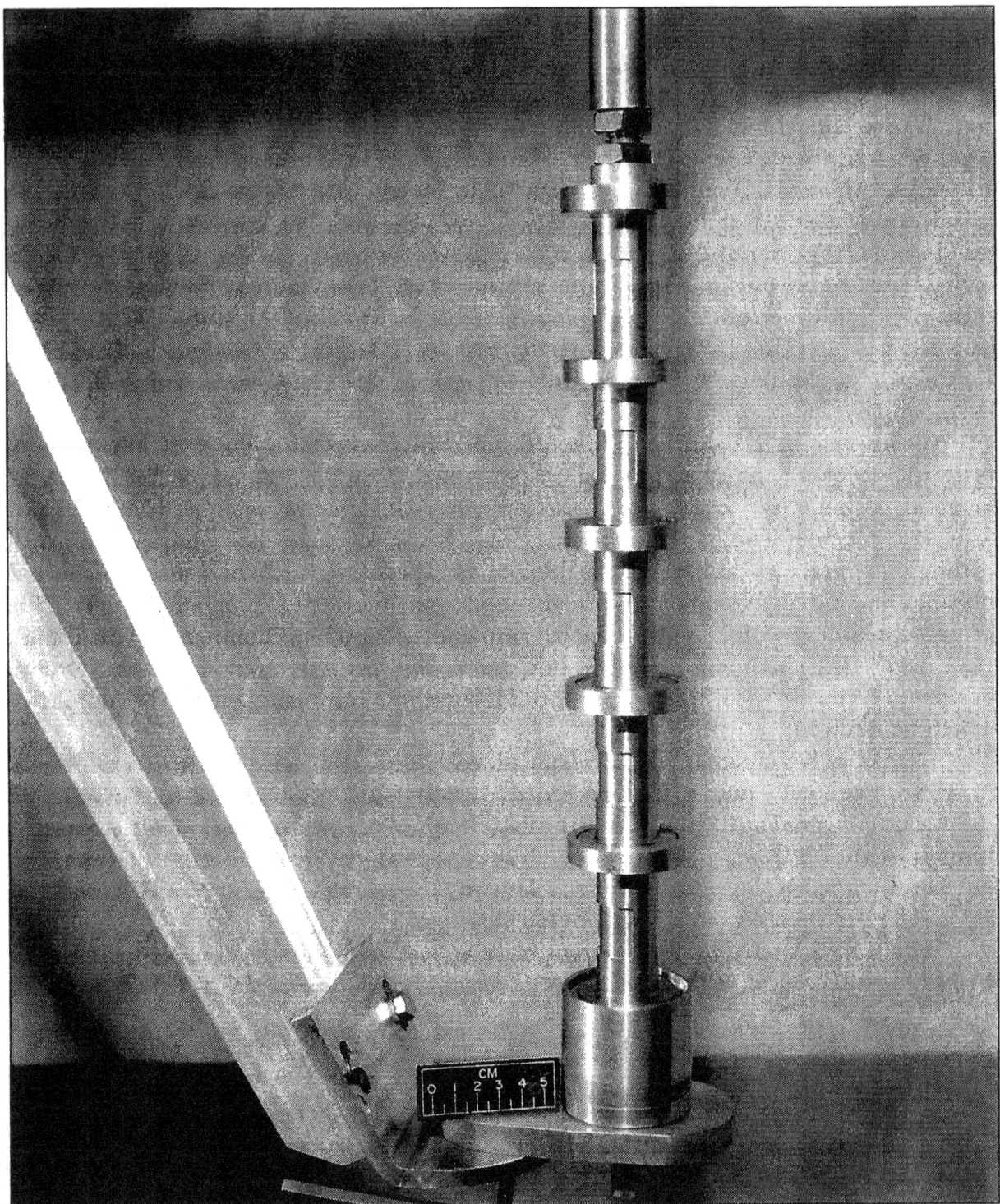

Figure 5–9. Ice-phobic test rig designed by Raymond Sotos, 1980. (NASA C-80-685)

idea, allowing needed work to be done in a variety of new areas. Aircraft icing research, everyone knew, was severely hampered by the limitations of ground icing simulation facilities such as the Canadian spray rig. While the IRT was the largest refrigerated icing tunnel in the world, it lacked the size to test many full-scale components that required icing protection. Also, it was limited to speeds under 300 miles per hour at sea level. Conversion of the AWT, which had not be used since 1970, would allow more accurate studies to be made of icing problems that were encountered by high-speed fixed- and rotor-wing aircraft.[15]

The appointment of a new director for the Lewis laboratory in 1982 seemed to offer an opportunity to persuade Headquarters to authorize the conversion. After Lewis planners recommended to Andrew J. Stofan that he use the usual "honeymoon" period accorded to new directors to pursue five major programs, including refurbishment of the AWT, Stofan selected Brent Miller to form an AWT project office. Reinmann, in turn, was told to add an icing group to the planning office.[16]

Over the next three years, the scope of the project grew significantly. It was difficult to "sell" the conversion of the AWT into an icing research facility when it would cost $10 to $20 million to support a program with an annual budget of $1 to $2 million. Under the circumstances, it was not surprising that the project office began to emphasize the many uses to which a refurbished AWT could be put. By 1985, planners were arguing that the AWT conversion would provide the aeronautical industry with "a needed, truly unique, and diverse wind tunnel for future propulsion system integration and icing R & D." Endorsed by the Department the Defense, the FAA, and industry, the new facility would "play a major role in maintaining U.S. superiority in aeronautics well into the 21st century."[17]

It all sounded fabulous. The cost for the conversion, however, had also become fabulous. The original $10 to $20 had risen to $150 million—and was growing. As Beheim had warned earlier, the decision to go ahead with the renovations would be "highly dependent on budgetary restraints." He was proved correct; the project priced itself out of existence.[18]

The three years had not been wasted for the icing group in the project office. A great deal had been learned about icing systems for wind tunnels. Researchers William Olsen

---

[15] B. J. Blaha and R. J. Shaw, "The NASA Altitude Wind Tunnel—Its Role in Advanced Icing Research and Development," a paper presented at the 23rd Aerospace Sciences Meeting of the AIAA, Reno, Nevada, 14–17 January 1985.

[16] Dawson, Engines and Innovation, pp. 213–14.

[17] Blaha and Shaw, "The NASA Altitude Wind Tunnel."

[18] Reinmann interview.

Figure 5–10. Raymond Sotos (left) and David Justavick during U.S. Air Force ice-phobic tests, 1980. (NASA C-80-2923)

and James Newton, for example, had undertaken a major study to find out where the water went after it left the spray nozzles and passed through the test section as an icing cloud. By collecting and weighing the ice that froze on the turning vanes, fan blades, and between the fins of the heat exchanger, Olsen and Newton were able to determine where and how much water was deposited at various points along the tunnel loop. "Nothing like this had ever been attempted before," Reinmann pointed out. "This was a very physically difficult job, and I always considered it a highlight of the history of our icing program."[19]

Although not prepared to spend the large amount of money to refurbish the AWT, NASA Headquarters did approve a major modification program for the heavily used IRT. In February 1986, the tunnel was shut down. It would not reopen until January 1988. Manager Zager was in charge of this modernization project. Under his watchful eye, improvements were made in the following six major areas:

1. New spray nozzles and an aerodynamically designed spray bar that had been developed for the AWT were placed in the IRT. Instead of the previous six spray bars and seventy-seven nozzles, the new arrangement featured eight bars. In addition to the standard air-assisted nozzles with a tube diameter of 0.025 inches, the new system also included a modified nozzle (mod-1) that had been designed by Olsen and featured a tube diameter of 0.0155 inches. When using standard nozzles, the bars contained ninety-four nozzles. With the mod-1 configuration, an additional nozzle was added. At 145 miles per hour, the standard nozzles produced a uniform icing cloud that measured 3 feet by 4 feet. With the mod-1 nozzles, the size of the cloud was 2 feet by 3 feet, but it more closely approximated the lower liquid content of natural icing clouds. Cloud uniformity varied with airspeed. At 70 miles per hour with standard nozzles, the cloud was uniform over much of the test area; at 220 miles per hour, the cloud shrank to 1.5 feet to 2.5 feet.

2. A new 5,000-hp drive motor replaced the motor that had been installed in 1944.

3. Most noticeable of all the innovations was a Westinghouse Distributed Process Control System, which provided programmable, digital control of the drive motor, refrigeration system, spray bar system, and other support systems. Prior to this, the IRT had used analog controls with a huge bank of manometers and paper recorders. Water came into the tiny control room, where flow rates and pressures were set by flow control valves. Three to four people crowded into the room to take data and to photograph the manometer board. It was a laborious, time-consuming process.

The computerized control system, the first time such a system was used at Lewis, brought the IRT up to state of the art. Operators now controlled water and air pressure, wind speed, and temperature by pressing keys on a keyboard, while they observed con-

---

[19] Reinmann to Leary, 1 June 2001.

Figure 5–11. Raymond Sotos views an A-10 inlet in the IRT, 1981. (NASA C-81-177)

ditions in the tunnel on a computer screen. Whereas it was possible to take only about 20 data points per hour (excluding manometers) with the analog system, researchers could now obtain up to 250 bits of information about the IRT and the model each second and display the data or record it for later analysis. Zager, however, retained a manometer board in the back room of the new control room to watch the compressor profile. He was apprehensive about the effect on the new system of a sudden loss of power. If not careful in handling the airflow when starting up and shutting down the tunnel, the compressor could wipe out the blades. (The manometer board was promptly removed upon Zager's retirement.)[20]

4. The new computerized control system was housed in a room that not only was three times larger than the old room, but it also featured vastly improved acoustics.

5. All wooden floors were replaced with concrete flooring.

After construction was completed, the tunnel underwent nearly seven months of calibration. This was a crucial process, as the ability of the IRT to simulate flight through natural icing clouds depended upon the calibration of airspeed, temperature, turbulence level, liquid water content, and droplet size. Calibration of the spray nozzles increased the upper limit of mean effective drop diameter from 20 to 40 microns, thereby increasing the range of natural icing conditions that could be simulated. By the time everything was finished, NASA had the finest icing research facility in the world.[21]

NASA's refurbished tunnel also served as a model in many ways for two new icing research tunnels. BFGoodrich Aerospace had dismantled its original icing tunnel in the 1970s. In 1986, the company decided to build a new tunnel for its Ice Protection Systems Division at Uniontown, Ohio. Goodrich wanted to test and develop new ice-protection products, especially the Pneumatic Impulse Ice Protection System (PIIP). The goals for this advanced system, which employed a 10-millisecond pulsed inflation to distort the surface of the boot momentarily, were to minimize boot intrusion into the airstream, require less ice thickness for removal, and better resist erosion.

"We called our friends at NASA Lewis to help," facility manager David Sweet recalled. "They openly shared with us the icing spray technology used in the IRT that allowed them to best replicate a natural icing cloud." Completed in 1988, the Goodrich tunnel had a subsonic, closed-loop configuration, with a refrigeration system in the center of the loop that could lower temperatures to −25°F. The test section, 5 feet in

---

[20] Zager interview.

[21] John J. Reinmann, Robert J. Shaw, and Richard J. Ranaudo, "NASA's Program on Icing Research and Technology," NASA TM-101989 (1989), first delivered as a paper for the Symposium on Flight in Adverse Environmental Conditions, Sponsored by the Flight Mechanics Panel of AGARD, Gol, Norway, 8–12 May 1989.

Figure 5–12. Electro-impulse de-icing equipment for McDonnell Douglas tests in the IRT, 1981. (NASA C-81-6007)

length, measured 44 inches high and 22 inches wide. An 80-inch-diameter fan, driven by a 200-horsepower rotary motor, produced airspeeds of up to 210 miles per hour. Five spray bars used twenty NASA-designed standard and mod-1 nozzles.[22]

NASA technology also assisted in the construction of a new Boeing icing tunnel. The existing tunnel at Boeing had an 18-inch by 30-inch test section and was used only for development work. With an upcoming program for the advanced 777 transport, Boeing decided that the company could save a good deal of money by building a larger facility to help with the airplane's icing certification in 1986. George E. Cain was responsible for designing and building a new tunnel with 4-foot by 6-foot and 5-foot by 8-foot test sections. And the job had to be done within three years. "My first action," he

---

[22] David Sweet to Leary, 24 May 2000.

recalled, "was to tour icing tunnel facilities in the U.S. and see what was out there and where their problems were."

Cain contacted Zager and arranged to visit the IRT. He spent several days at the facility, "finding out what their problems were and what would they change if they were going to build a new tunnel." As a result of his discussions with Zager, Cain included in the design concept for the Boeing Research Aerodynamic Icing Tunnel (BRAIT) significant modifications of the heat exchanger and spray bar system. The BRAIT facility, which opened in 1989, saw improvements on the Lewis facility, especially in the areas of nozzle control and airflow. Boeing would later share these advances with NASA. "Ever since the beginning of our relationship with NASA," Cain observed, "we have maintained an open-door policy to discuss operational problems with each of our facilities . . . . We have an excellent relationship with NASA."[23]

While NASA's new tunnel was taking shape, the old tunnel was being recognized for its contribution to icing research by the American Society of Mechanical Engineers (ASME). In 1985, Robert Graham, NASA Lewis liaison with the ASME, had suggested to Reinmann that the IRT might be a good candidate for designation as a historic engineering landmark. At the time, Reinmann had taken no action on Graham's proposal and stuffed the ASME brochure on engineering landmarks in his desk drawer. Two years later, however, after noting the extensive historical work that William Olsen had done in support of the AWT modification program, especially on the Carrier system, Reinmann pulled out the ASME brochure, handed it to Olsen, and suggested that he undertake the project. Olsen was delighted with the assignment. "Bill," Reinmann recalled, "poured his heart into this."[24]

Assisted by Virginia P. Dawson, a contract historian, Olsen put together a package of documents and photographs in support of the nomination. He listed three reasons why the IRT should be considered historically significant. First, he observed that most of the current aircraft icing technology had been researched, developed, and tested in the IRT. Second, the IRT was the largest refrigerated icing tunnel in the world. Finally, at 21,000 tons capacity, the refrigeration plant used by the IRT remained the largest direct expansion refrigeration system in the world. The heat exchanger in the IRT featured a revolutionary design that Willis Carrier considered to be his greatest engineering achievement. Olsen also included a diagram of the IRT's innovative spray nozzles.[25]

The ASME accepted Olsen's nomination and agreed to add the IRT to such other landmarks as the Five-Foot Wind Tunnel at Wright Field, the Link Trainer, and the

---

[23] George E. Cain to Leary, 3 November 2000.

[24] Reinmann to Leary, 1 June 2001.

[25] Olsen and Dawson to ASME, "Nomination for Designation of Historical Engineering Landmark," n.d., NASA History Office.

Sikorsky VS-300 Helicopter. The ceremony took place at Lewis on 20 May 1987. Many of the individuals who were responsible for the design and construction of the IRT were present, along with the researchers who had worked in the facility during the 1940s and 1950s. The ASME formally designated the IRT as the 21st International Historic Mechanical Engineering Landmark and placed a plaque on the building that proclaimed the following:

> *The icing research tunnel is the oldest and largest refrigerated icing wind tunnel in the world. The world's aircraft fly safely through icing clouds because of the technology developed in this tunnel. Two important achievements of the installation are the unique heat exchanger and the spray system that simulates a natural icing cloud of tiny droplets.*[26]

---

[26] Program, "An International Historic Mechanical Engineering Landmark: Icing Research Tunnel," 20 May 1987, History Office, GRC.

Chapter 6
Full Speed Ahead

The improvements in the IRT, Reinmann observed shortly after the tunnel reopened in 1988, not only increased the productivity of the facility, but they also provided new test capabilities. The need for these new capabilities was soon demonstrated when NASA agreed to conduct a joint test program in the IRT with the Boeing Airplane Company to evaluate ground de-icing fluids. The tests would require a rapid increase in airspeed to simulate takeoff, which could not have been done with the old drive motor and controls.

The Air Florida accident in 1982 had focused attention on the largely neglected subject of ground de-icing fluids. How long did these fluids last under a variety of weather conditions? How did they affect aircraft performance, stability, and control? It turned out that there were few reliable answers to these questions. In North America, de-icing had been accomplished by spraying aircraft with a mixture of ethylene glycol and hot water. This solution lowered the freezing point of precipitation in both its liquid and crystal (ice) phases. The glycol solution was economical and easily applied; however, it provided protection for only a limited time and had to be frequently reapplied. In Europe, the Association of European Airlines (AEA) used a similar fluid, known at AEA Type I, and a thickened mixture, which was designated AEA Type II.[1]

In the days after the Air Florida accident, U.S. airlines began to take an interest in European de-icing fluids, especially Type IIs. Using a polymer thickener, Type IIs had the consistency of a gel when there was no air flowing over the surface; with airflow, such as during takeoff, the shearing action reduced the viscosity appreciably. They provided much longer protection than the Type Is, but information on the effect of these fluids on takeoff performance remained sketchy.

Responding to airline concerns about the possible aerodynamic effects of the AEA fluids, Boeing performed a series of tests in its small-scale wind tunnel, using 10-inch and

---

[1] Reinmann, "Icing: Accretion, Detection, Protection," in Advisory Group for Aerospace Research & Development (AGARD), Lecture Series 197: Flight in an Adverse Environment (NATO, 1994), pp. 4.1 to 4.27.

Figure 6–1. The electro-impulse de-icing system inside a wing during IRT tests, 1981. (NASA C-81-6083)

24-inch airfoils in 1982. Results of this investigation indicated that both the undiluted Type I and Type II fluids failed to completely flow off the wing prior to reaching liftoff velocity. The remaining residue, the researchers noted, formed a rough, reticulated surface, causing a measurable loss in lift and an increase in drag.[2]

Concerned about the results of the Boeing tests, the AEA sponsored a research program in 1984 with the von Karman Institute for Fluid Dynamics of Brussels, Belgium, to further investigate the possibility of aerodynamic penalties caused by the fluids. Using a 4.92-foot-chord two-dimensional model, the Institute conducted a series of experiments on the effect of the fluids on takeoff performance during subfreezing temperatures. The results of these tests generally agreed with Boeing's findings. In the wake of this troubling conclusion, AEA asked Boeing to perform a flight test investigation of the fluid effects.

Before the flight tests could take place, however, two major accidents made the dangers of improper de-icing dramatically clear. At 5:34 A.M. on 12 December 1985, an Arrow Air DC-8 landed at Gander, Newfoundland, for refueling on a flight with U.S. military personnel from Europe to the United States. Although light freezing drizzle was falling during the refueling process, the pilot did not request de-icing. Taking off at 6:45 A.M., the aircraft pitched up after rotation, banked to the right, and struck the ground 3,000 feet beyond the end of the runway. The crash killed all 248 passengers and eight crew members. Although the investigation of the accident took place amidst controversy about the possibility that the plane had been brought down by a bomb, a majority of the Canadian investigating board and a judicial review concluded that the DC-8 had stalled due to ice contamination of the leading edges and upper surfaces of the wing.[3]

A second major accident took place on 15 November 1987. Moderate snow was falling as a Continental DC-9 prepared to depart Denver. The airplane was de-iced, but 27 minutes elapsed before the flight received takeoff clearance. Although company procedures called for additional de-icing if delays exceeded 20 minutes, the pilot elected to continue. The results were similar to the Arrow Air crash. The DC-9 stalled after rotation and impacted upon the right side of the runway. Twenty-eight people lost their lives.[4]

---

[2] The work on de-icing fluids is summarized in Reinmann, Shaw, and Ranaudo, "NASA's Program on Icing Research and Technology," NASA TM 101989 (1989), and Eugene G. Hill and Thomas A. Zierton, "Flight and Wind Tunnel Tests of the Aerodynamic Effects of Aircraft Ground Deicing/Anti-Icing Fluids," a paper presented at the 29th Aerospace Sciences Meeting of the American Institute of Aeronautics and Astronautics, Reno, Nevada, 7–10 January 1991.

[3] See *New York Times*, 9 December 1988; *Toronto Star*, 15 March 1989, and 19 July 1989.

[4] For an excellent summary of icing-related accidents between 1946 and 1996, see the special issue of *Flight Safety Digest*, "Protection Against Icing: A Comprehensive Overview," June/September 1997.

Figure 6–2. IRT spray bars, 1982. (NASA C-82-4308)

The Boeing/AEA flight tests were conducted at Kuopio, Finland, between 11 and 20 January 1988. Lufthansa Airlines provided a Boeing 737-200, while Boeing instrumented the aircraft and performed the tests. Four undiluted de-icing/anti-icing fluids were used: an AEA Type I that was manufactured by Hoechst, A.G.; an obsolete Type II fluid for which wind tunnel data was available; a Hoechst Type II fluid that was in current airline use; and a Type II fluid provided by Union Carbide Canada, Ltd. Objectives of the flight tests included an evaluation of the aerodynamic effects of the four fluids and the accumulation of a database for extrapolating data from a more comprehensive wind tunnel investigation in the future.

Equipped with a data-recording system that monitored gross weight, center of gravity, engine parameters, airspeed, and acceleration, the Boeing 737 made a series of takeoffs in various configurations as video and photographic records were made of the four fluids, which had been treated with fluorescent dye to improve visibility. Researchers were able to measure fluid depth and roughness by using ultraviolet photographic and back-scatter laser techniques. Prevailing ambient temperature permitted data to be obtained over a range of 36° to 14°F. The tests confirmed the 1982 Boeing tunnel results with all fluids causing a measurable loss in lift. At lower temperatures, the lift loss was greater with the Type II fluids. Also, losses were larger at a flaps 15° takeoff configuration than at flaps 5°.

NASA became involved in the evaluation of ground de-icing fluids in the spring of 1988. Richard Adams, who had left the Army to become the national resource specialist on aircraft icing for the FAA, was largely responsible for bringing Lewis's icing branch into the picture. Adams was the FAA's point-of-contact for ground de-icing issues and had worked to coordinate the efforts of the SAE, AIAA, Air Transport Association, Air Line Pilots Association, American and foreign aircraft manufacturers, and the AEA. At this time, the FAA did not have a procedure for approving specific de-icing fluids. De-icing procedures came under Operations Specifications, which required aircraft manufacturers to assure the airlines that the fluids were compatible with the materials on which they were to be applied; however, the manufacturers did not address the question of aerodynamic effects of the fluids. Adams sought reliable data on which to base new aerodynamic standards that would govern the use of de-icing fluids.[5]

Adams arranged with Reinmann for the use of the IRT following the Boeing 737 flight tests. The experiments would be a joint effort by NASA, Boeing, the AEA, and various fluid manufacturers. Boeing would provide the wind tunnel models and conduct the tests; NASA would operate the IRT; and AEA observers would monitor the process. The first phase of the tests, which took place in April 1988, was designed primarily to assess

---

[5] Adams to Leary, 9 November 2000.

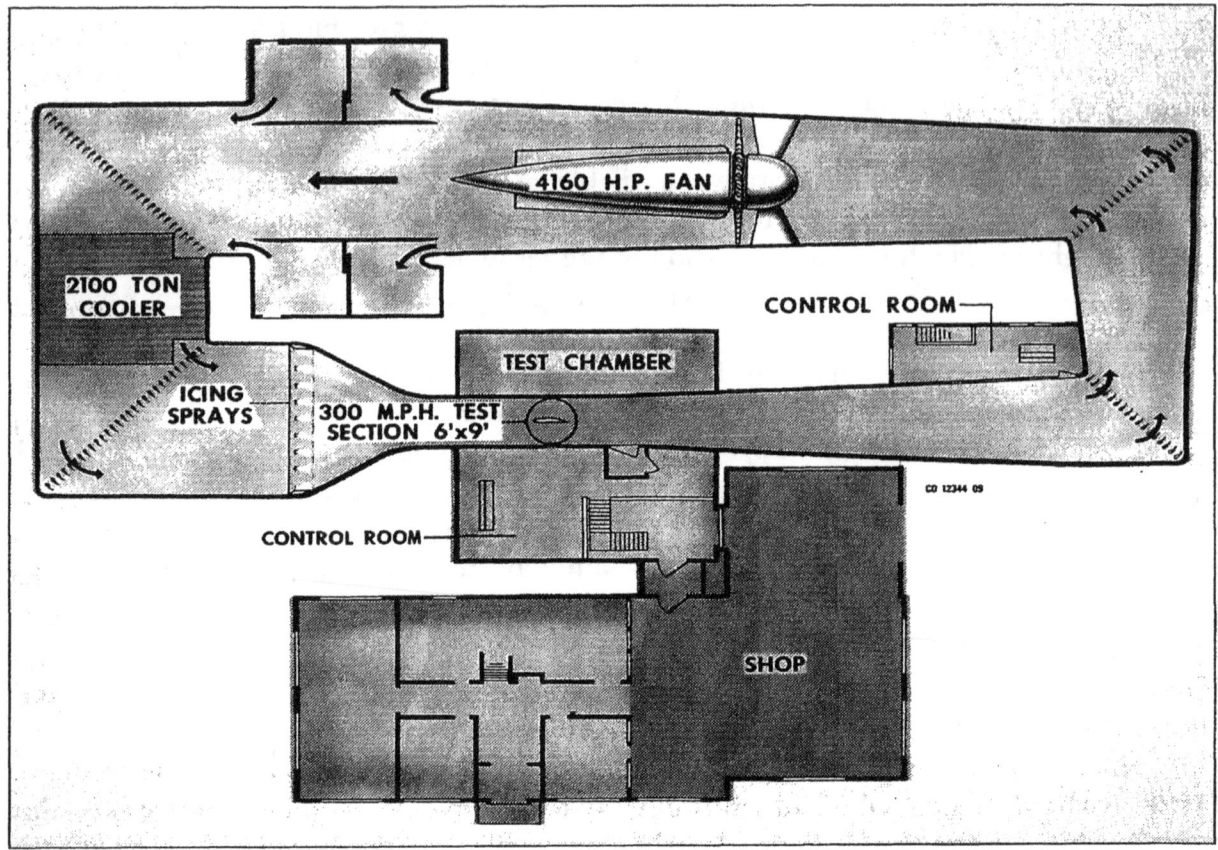

Figure 6–3. Schematic of the IRT, 1983. (NASA C-83-870)

the correlation between the flight results and wind tunnel data. Using the same de-icing fluids on a 737-200 three-dimensional model, Boeing researchers Thomas A. Zierten and Eugene C. Hill found that results were sufficiently similar that wind tunnel lift loss results could be applied to full scale on a 1:1 basis. The tests also revealed that the aerodynamic effects of the fluids were strongly influenced by the aircraft's operational speeds and time to liftoff.

The second phase of IRT testing was done in February 1990. Boeing researchers Zierten and Hill investigated the aerodynamic effects of three Type I and four Type II fluids that were currently in production, plus eight fluids that were under development by manufacturers. By the time the tests were completed, additional data had been secured to permit the establishment of an aerodynamic test standard for both Type I and Type II fluids.

Adams ensured that the information that was being acquired on de-icing received wide publicity, primarily through a series of FAA-sponsored conferences. The adoption of standards was made urgent following the crash of a USAir Fokker

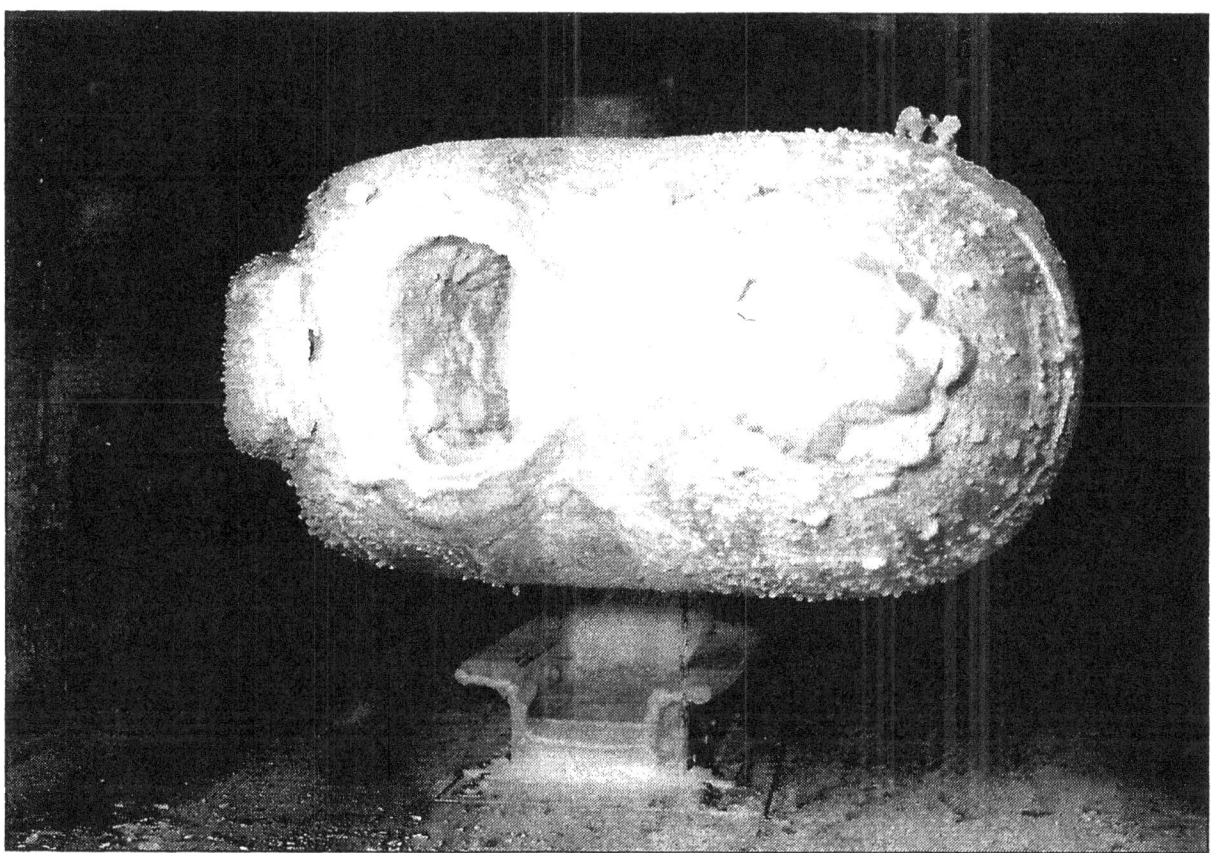

Figure 6–4. Fairchild SF340 in IRT, 1983. (NASA C-83-1160)

F-28 on 22 March 1992. The aircraft had been de-iced twice before leaving the gate at New York's LaGuardia Airport, but 35 minutes had elapsed between the last de-icing and takeoff. The Fokker stalled after rotation and came to rest in Flushing Bay (at the end of the runway). There were twenty-seven fatalities in an accident that was given extensive publicity.[6]

---

[6] *New York Times*, 23 March 1992. Following this accident, Reinmann was asked to appear on a local TV show in Cleveland, together with a survivor of the crash. "I used some graphics and a model airplane to illustrate how ice buildup on the wings degrades aerodynamic performance and the consequences," he recalled. "The next day, a captain from one of the major airlines called me from California, his home base. He said that he had watched the show while he was on a holdover in Cleveland, and he wanted me to know that it was the first time that he really understood how icing affected takeoff performance. He said that when he got on the airplane he mentioned the show to his copilot, and he too had watched it and found it helpful. The captain talked to me for about an hour, telling me about his experiences in icing." Reinmann to Leary, 22 April 2001.

Figure 6–5. Icing section in 1985. Standing: (L/R) John Reinmann, Nikola Juhasz, Andrew Reehorst, Sandy Kosakowski, Robert Ide, Gary Ruff, Peggy Evanich, Kevin Mikkelson, James Newton. Kneeling: (L/R) Mark Potapczuk, Thomas Miller, Robert Shaw, Gary Horsham, William Olsen, Kevin Leffler. (NASA C-85-3120)

While the NTSB faulted the pilot for early rotation, the accident investigators levied their major criticism at the FAA. The probable cause of the accident, the NTSB concluded, was the failure of the FAA and the airline industry to provide flight crews with procedures, requirements, and criteria for de-icing when facing delayed departures.[7]

In the wake of the USAir accident, the FAA moved rapidly to formulate new rules on aircraft icing before the winter of 1992–93. Effective 1 November, the new regulations specified the maximum allowable time between de-icing and takeoff. Additional information followed on the characteristics of various de-icing fluids, largely based on the testing that had been done between 1982 and 1990.[8]

---

[7] *Washington Post*, 21 September 1992.

[8] FAA Press Release, 25 September 1992.

Figure 6–6. Sikorsky PFM awaiting installation in the IRT, 1986. (NASA C-86-4570)

Reinmann took considerable pleasure in NASA's contribution to the development of ground de-icing standards. The modernized IRT had performed superbly, allowing the results of scale-model tests to be extrapolated to full size. Even Boeing researcher Zierten seemed impressed and told Reinmann that the IRT's productivity was at least as high as that of the Seattle manufacturer. Years later, when asked about the major accomplishments during his sixteen years as head of icing research, Reinmann placed the work on ground de-icing near the top of his list.[9]

NASA also participated in a joint program with McDonnell Douglas Aerospace (MDA) to study the effect of icing on the performance of advanced, high-lift airfoils. In order to compete with Airbus, U.S. manufacturers were working on a simplified three-element wing design that would improve performance and consume less fuel. Early in

---

[9] Reinmann interview.

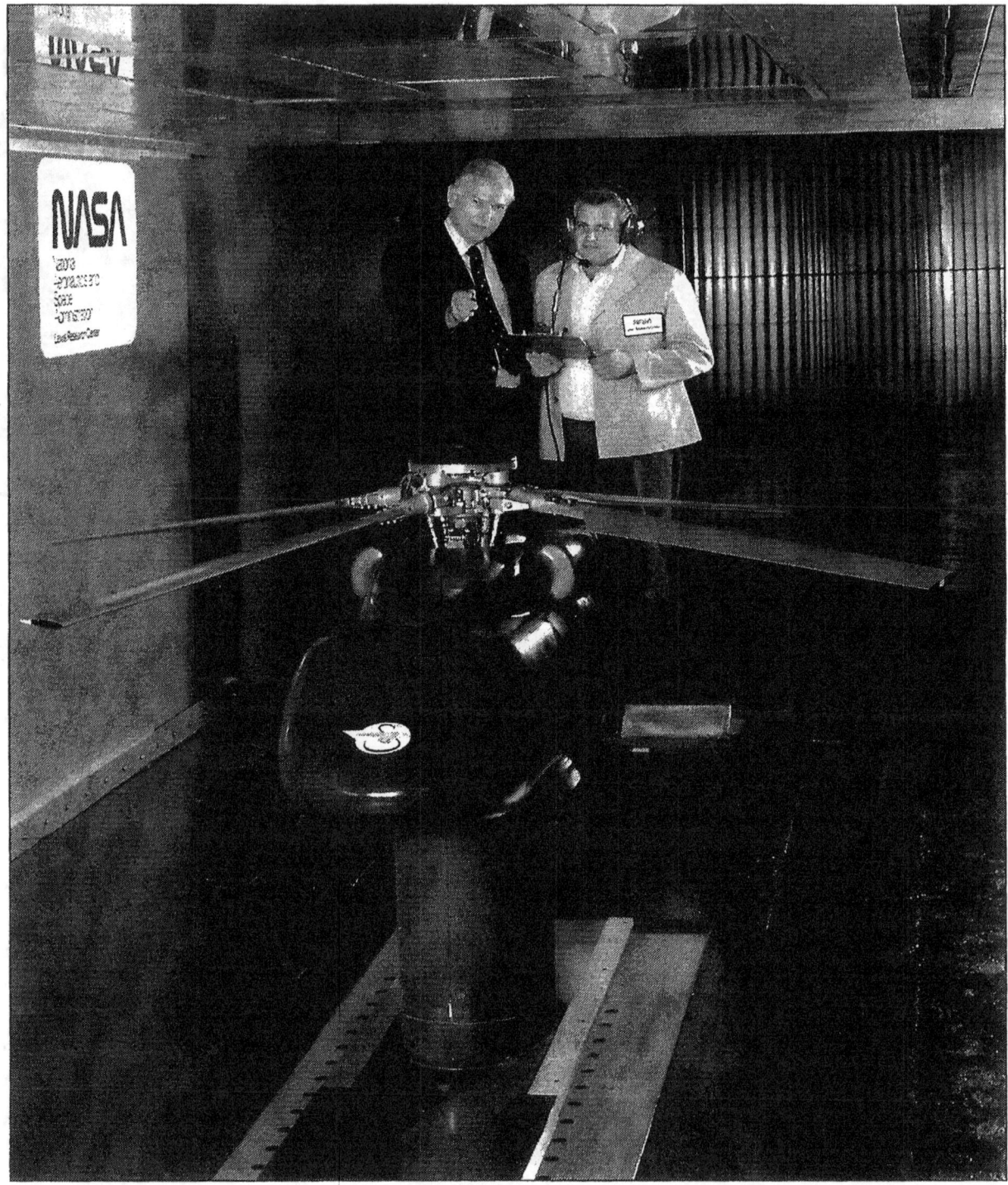

Figure 6–7. John Reinmann (L) and Thomas Bond with Sikorsky PFM in the IRT, 1989. (NASA C-1989-9699)

1992 when NASA and MDA discussed possible icing programs of mutual interest, the three-element wing design was a natural candidate for investigation.[10]

MDA designed and fabricated a model of the three-element wing specifically for vertical installation in the IRT. It consisted of a slat with a hot-air de-icing system, a main element assembly, and a large-chord single-segment flap. Tests began in the summer of 1993 under the direction of project leader Jaiwon Shin. They revealed that the slat reacted to icing much like the leading edge of a single-element airfoil, whereas the main element assembly and flap were much less sensitive. Also, the entire lower surface of the flap accreted ice under all icing conditions, with significant buildups on the trailing edge in most cases. Finally, the gaps between the elements did not suffer any ice contamination for the icing conditions and flap settings used in the tests.[11]

Alan Kehlet, vice president for advanced program and technology/transport aircraft at MDA, considered the tests "very productive." The quality and amount of data that had been collected, he wrote to NASA, "is due in large part to the working relationship established between the NASA team and McDonnell Douglas personnel, and the quality support received from the NASA engineers and technicians involved in this test."[12]

Additional testing took place in the summer of 1994 to add to the database on the three-element system and to explore at greater length the possible effects of gap sizes. In some 177 icing runs, Shin and his team found that slat-ice accretion was more sensitive to angle-of-attack changes than the main element or flap. Also, changing flap settings had a smaller effect on ice accretion than did changing the angle of attack. Work in the IRT ended at this point; although, additional experiments were performed in Langley's Low-Turbulence Pressure Tunnel using ice shapes that had been generated during the IRT tests.[13]

NASA's cooperation with McDonnell Douglas and Boeing were the types of developmental programs that generated a good deal of favorable publicity for the space agency. Unlike the pre-1958 era, the post-1978 years saw far more time devoted to this kind of program than to more basic research. By early 1990s, with the IRT in use over 1,000

---

[10] Reinmann, "Cooperative NASA/Industry/Agency/University Programs," NASA-Industry Workshop on Aircraft Icing, NASA Lewis Research Center, 27–29 July 1993.

[11] Dean Miller, Jaiwon Shin, and David Sheldon (NASA); Abdolah Khodadoust and Peter Wilcox (MDA); and Tammy Langhals (NYMA, Inc.), "Further Investigations of Icing Effects on an Advanced High-Lift Multi-Element Airfoil," NASA TM 106947 (1995).

[12] *Lewis News*, 12 August 1994.

[13] Ibid.; Miller, et al., NASA TM 106947.

Figure 6–8. NASA's Twin Otter behind Army's HISS, 1988. (NASA C-88-1728)

hours a year, approximately 60 percent of tunnel time was being spent in direct support of civilian and military manufacturers.[14]

Reinmann, however, also ensured that his section made a strong effort to better understand the phenomena of icing and to develop better tools to predict its effects. The advent of high-speed computers, for example, opened up the possibility of developing a computer program to predict ice-accretion shapes, design aircraft de-icing systems, and analyze performance degradation due to icing. The increasing cost of flight tests for icing certification certainly provided strong motivation to employ aircraft icing analyses methods wherever possible. Reinmann assigned primary responsibility for this area of research to his deputy, Robert J. Shaw. "I gave him a lot of freedom," Reinmann recalled, "and he loved it."[15]

---

[14] Les Dorr, Jr., "We Freeze to Please . . .," *NASA Magazine*, spring 1993, pp. 26–30.

[15] Reinmann interview.

Figure 6–9. Boeing 737 model in the IRT during tests of ground de-icing fluids, 1980. (NASA C-1990-7099)

Shaw learned that the University of Dayton was doing modeling on the effect of heavy rain on airfoil performance. It would be a logical extension to model icing. Shaw contacted Professor James K. Luers, director of the University of Dayton Research Institute (UDRI), in 1980 and arranged for a grant from NASA to develop an airfoil ice-accretion model. The model, Luers recalled, used "a control volume approach to calculate the physics and thermodynamics of super-cooled water droplets impacting an airfoil surface." Charles D. MacArthur, who was primarily responsible for performing the technical work on the codes, dubbed them LEWICE.[16]

Development of the codes, Shaw emphasized, involved "terribly complicated physics." The team at UDRI worked for more than two years to improve and extend the model so that it provided reasonably good predictive capability for complex ice-accretion

---

[16] Luers to Leary, 26 July 2001.

117

Figure 6–10. Imaging equipment in the IRT control room, 1990. (NASA C-1990-6833)

shapes under both rime and glaze icing conditions. LEWICE was then delivered to NASA Lewis, where it underwent further development and extensions. LEWICE 1.0 was formally released to U.S. industry in May 1990.[17]

By the early 1990s, LEWICE codes had been validated through a wide range of conditions and were being used routinely by aircraft manufacturers during the certification process. For example, Richard Ranaudo, an experienced NASA test pilot, retired in 1998 to become manager of Canadair Flight Test at Bombardier Aerospace in Wichita, Kansas, and lead three certification and development programs for the Canadair Regional Jet Series. "Since joining 'industry,'" he has written, "I've learned that NASA's icing program in computational methodologies for ice shapes [has] been the biggest help in new aircraft development."[18]

---

[17] Leary interview with Robert J. Shaw, 26 June 2001.

[18] Ranaudo to Leary, May 2001.

Work on LEWICE continued during the 1990s, with releases of versions 1.6 in June 1995 and 2.0 in February 1999. The codes, Shaw noted, work well in certain areas but not in warm temperature icing, due to the icing physics associated with those conditions. But the future of computer simulations remains bright. The development of three-dimensional codes has brought new challenges to scientists—and promises great rewards.[19]

Validation of computer codes during the 1980s and 1990s involved flight tests and experiments in the IRT. Reinmann appreciated the value of flight testing and lent his support to the work of the group, although not as much as the section sometimes desired. For one thing, flight testing was expensive, costing about one-third of the icing branch's $1 to $2 million annual budget. Also, Reinmann noted, the flight group usually wanted to build a program around the airplane, whereas Reinmann wanted to use it as part of a broader icing effort, such as the validation of computer codes, testing advanced ice-protection concepts, and assessing the effects of ice on airplane performance.

Certainly, the Twin Otter that was used for icing research had proved an extremely useful tool for the work of the icing section. The airplane was equipped with electro-thermal anti-icers on its propellers, engine inlets, and windshields. It had non-standard pneumatic boots on the vertical stabilizer, wing struts, and landing struts, and standard boots on wings and engine nacelles. Researchers measured leading edge ice shapes with a stereo photographic system. A wake survey probe mounted on the wing behind the area where the photos were taken recorded wing section drag, while a noseboom provided information on airspeed, angle of attack, and sideslip.

Averaging about thirty missions a year since 1981, the Twin Otter, usually flown by Ranaudo, had numerous encounters with natural icing clouds, during which ice was allowed to build up on the leading edge of the wing. The aircraft was then flown into clear air so that stereo photographs could be taken of the ice shapes and performance degradation measured. This inflight data was then used both to validate computer codes and to confirm that the IRT was adequately simulating natural icing.[20]

In addition to validating computer codes, the IRT also was used extensively in scale modeling programs, especially for helicopters. The rotorcraft industry was one of the fastest growing segments of civil aviation during the 1980s, and demands increased for all-weather capability. Helicopter companies, as noted earlier, had been using the IRT since the early 1960s to test engine inlets and ice-protection systems on stationary rotor blades, but full-scale main rotors could only be tested in actual icing conditions. With

---

[19] Shaw interview.

[20] For an example of this work, see R. J. Ranaudo, et al., "The Measurement of Aircraft Performance and Stability and Control After Flight Through Natural Icing Conditions," NASA TM 87265 (1986).

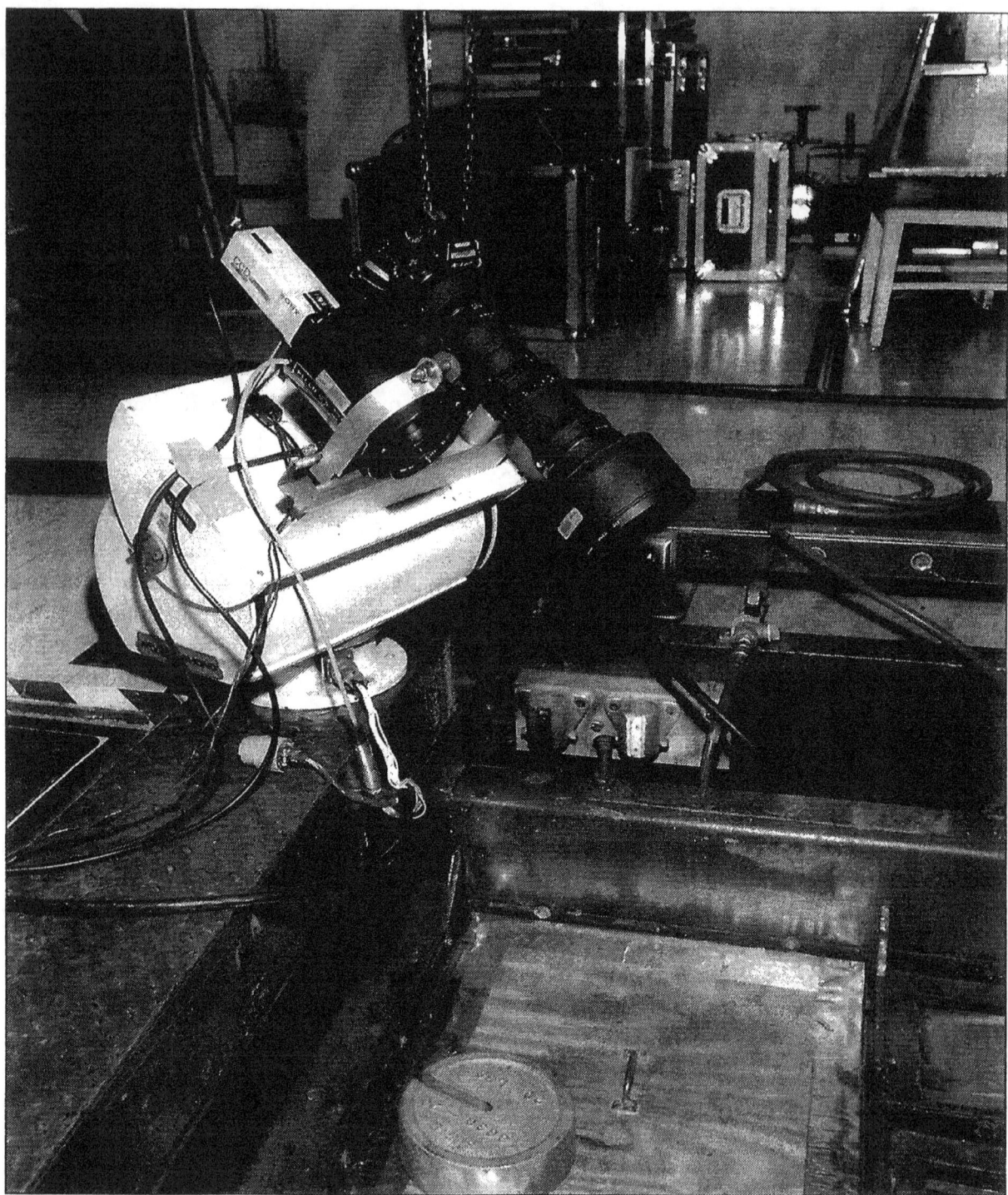

Figure 6–11. 35 mm still imaging camera with 400 mm Nikkor lens and video monitor attached to record experiments in the IRT. (NASA C-92-4309)

their short range and slow speed, helicopters usually had to wait until suitable icing conditions were available close to their home base; that is to say, they had to wait for the weather to come to them, whereas fixed-wing aircraft could fly to icing weather. Flight testing was both dangerous and expensive. And it could take years to acquire the data needed for FAA certification or military qualification. A possible answer to this problem would be the testing of a properly scaled model in the IRT.

In 1985, NASA and the Army sponsored the formation of the Rotor Icing Consortium, which included the four major helicopter manufacturers and Texas A & M University. The consortium selected a model one-sixth of the size of a Sikorsky UH-60 Blackhawk helicopter for testing in the IRT. Heavily instrumented, the Blackhawk was designated a Powered Force Model (PFM). NASA's responsibility would be to provide the facility and develop testing techniques.

Thomas H. Bond, the icing section's lead researcher for the project, suggested that the best way to explore rotor icing test procedures for the PFM would be to first test a simpler model rotor in the IRT. The Lewis laboratory, he told the consortium, had designed and built a rugged and heavy instrumented Bell OH-58 Kiowa tail rotor rig in 1980. The model, however, had never been tested and was in storage. He proposed that the rig be refurbished and serve as a test case for the PFM experiments. The consortium agreed.

The OH-58 rig was an ingenious creation of the Lewis model shop. It consisted of the tail shaft, hub, teetering components, and rotor blades for an OH-58. This assembly, 5 feet in diameter, was then mated to a NASA-designed drive force by a 40-horsepower electric motor. A 7-foot drive shaft allowed the rotor blades to turn in the middle of the IRT test section while the drive system remained underneath the tunnel turntable. A hydraulically operated center tube controlled the collective pitch of the rotor blades and enabled the blade angle to be changed quickly, if desired.[21]

Tests of the OH-58 rig took place in August, September, and November 1988. Concerned about the potential failure of the rotating hardware, Lewis technicians put together a set of armor plates on rails that was designed to protect vulnerable areas of the tunnel wall. Because these plates blocked the normal viewing area for tests, video cameras were installed to monitor the procedure and collect data.

As the tests began, the rig operator brought the tail rotor up to speed. Next, the tunnel's fan was turned on. Once tunnel conditions had stabilized, the operator adjusted the collective pitch to a 5°-forward tilt to simulate forward flight and started the data acquisition hardware. The spray was then introduced, the icing cloud formed, and the data was recorded on the rate of ice accretion. After the test run, photographs were taken of the resultant ice formation, the shape of the formation was traced, and ice molds were made.

---

[21] For a detailed description of the scale model helicopter experiments, see *Lewis News*, 30 March 1990.

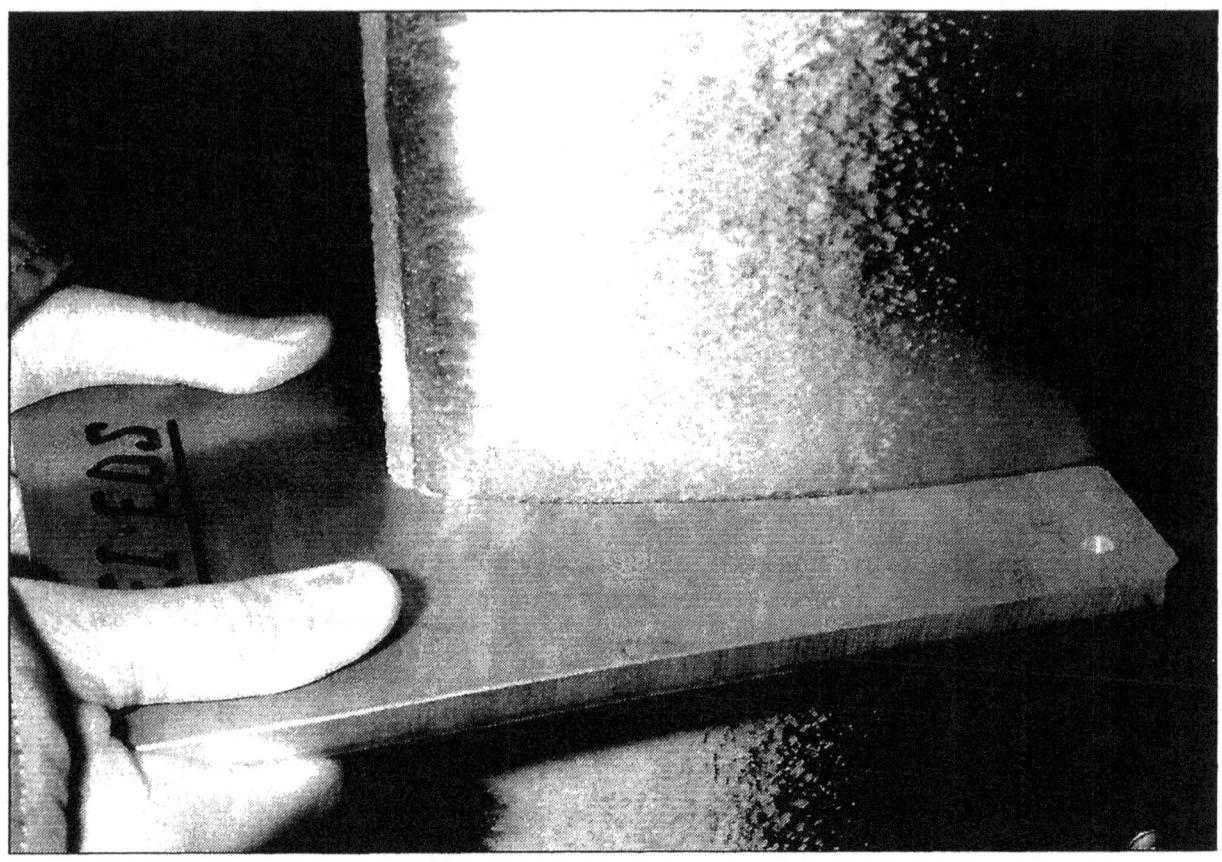

Figure 6–12. Metal template melting ice accretion for profile check, 1990. (NASA C-1990-13574)

The OH-58 tests proceeded without incident. As Bond had hoped, they paved the way for the PFM experiments. Lessons had been learned about coordination procedures for model and tunnel, and the best methods for observing and documenting rotor ice accretion and shedding. Also, the tests had led to establishing safety and emergency procedures.

PFM testing took place over a ten-week period in the fall of 1989. The Sikorsky-built Blackhawk model was a far more sophisticated creation than the OH-58 rig. It featured a six-component internal force balance from which measurements of main rotor performance could be made. The four-bladed main rotor consisted of generic NACA 0012 airfoils, 6 feet in diameter. (A true one-sixth-scale model of the Blackhawk main rotor would have been 9 feet in diameter, but the IRT test section could not accommodate a blade of this length.) With the PFM, a pilot sat in the control room of the IRT and "flew" the model with a joystick during icing tests. Bond collected performance data first with a clean model to establish a baseline, and then he gathered information under icing conditions.

The tests were promising, but it was clear that additional experiments would be necessary. A second series of tests were run in 1993. This time, the PFM was equipped with four 6-foot sections of SC 2110 main rotor blades—the same type of airfoil used on the Blackhawk. Exceptional model reliability, a faster data-acquisition system, and improved tunnel and model operational procedures led to high productivity. In 208 icing runs over a wide range of conditions, researchers obtained over 400 ice shape tracings and photographs. Also, they made eight sets of molds of ice accretion on the rotor blades. The experiments met Bond's highest expectations. Not only did they produce sizeable amounts of useful data, but they also demonstrated that a model rotor could be successfully tested in the IRT.[22]

Results of the PFM tests promised to supply major improvements in predicting rotorcraft performance in icing conditions. "The new experimental database resulting from these tests," Bond stated, "will be used to refine and validate a computer code that Sikorsky and NASA developed in earlier testing." The code could be used to predict full-scale helicopter performance in icing after it had been further validated in flight tests. Also, the molds constructed of ice shapes that had been formed on the PFM would be attached to scale-model rotors for tests in a conventional wind tunnel, and the results would be compared with IRT data. "We hope," Bond concluded, "that the code predictions and artificial ice shapes ultimately can be accepted as an alternative to some of the icing flight testing now required by the Federal Aviation Agency to certify a helicopter for flight into forecasted icing."[23]

It was during the PFM tests that Bond came to Reinmann's attention as a tenacious researcher. Even during routine testing, work in the IRT could be physically demanding. Researchers and technicians had to endure extreme cold temperatures and a noisy control room. Because the control room was operated at the same pressure as the test section, every time IRT speed increased, pressure in the test chamber dropped, causing their ears to pop in the control room. Not only was this annoying, but some people could not tolerate the continual increases and decreases in pressure. Helicopter rotor testing brought unique dangers. Although the walls of the test section had been armored, a shed blade could damage the Freon-filled heat exchangers. "A massive Freon leak," Reinmann pointed out, "could be catastrophic."

After several weeks of PFM testing at night and analyzing the data the next day, the research crew had become exhausted. At this point, a problem with the model developed. "I remember going over the IRT one evening," Reinmann recalled, "to discuss the possibility of terminating the test and sending the PFM back to Sikorsky to get it fixed. Tom and the

---

[22] Randall K. Britton (Sverdrup Technology), "Model Rotor Testing in the IRT," NASA-Industry Workshop on Aircraft Icing, NASA Lewis Research Center, 27–29 July 1993.

[23] NASA press release, 13 August 1993; Bond interview.

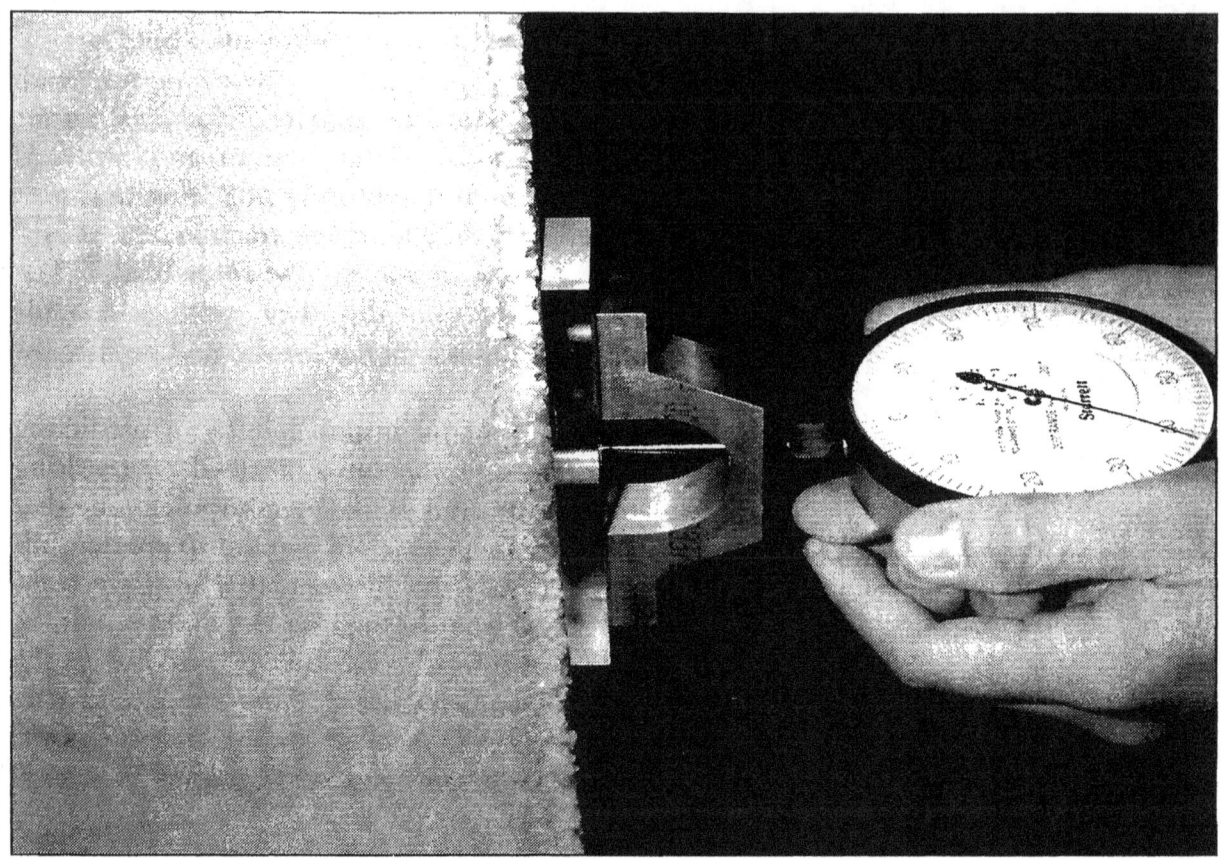

Figure 6–13. Measuring leading-edge thickness, 1990. (NASA C-1990-12947)

Sikorsky engineer didn't want to give up, in spite of their obvious exhaustion. Finally, I told them that we weren't at war, and they didn't have to work this hard. Besides, in their exhausted state, a serious mistake was more likely. So we terminated the test, but I was very impressed with Tom's determination to keep going in spite of his exhausted condition." The PFM went back to Sikorsky, where it was repaired. It returned to successfully complete the program.[24]

Bond also was the lead researcher in a joint project with the U.S. Air Force to assess the viability of recently developed low-power de-icer systems for engine inlets. This program began in 1990 after the Air Force's Oklahoma City Air Logistics Center asked NASA to investigate the new technologies. NASA agreed and advertised in the *Commerce Business Daily* that they were seeking manufacturers who wished to participate in the test program. NASA pointed out that, under the Space Act Agreement, no money would change hands. NASA would supply the test time in the IRT and an airfoil model, while the manufacturers

[24] Reinmann to Leary, 22 April 2001.

would be responsible for applying the ice-protection system on the model. Five manufacturers signed the agreement and supplied eight systems for evaluation.[25]

The tests took place in the summer of 1990 under the direction of Bond and Jaiwon Shinn of the icing branch, and Geert A. Mesander of the Air Force. The eight systems that were investigated were based on variations of three different technologies. BFGoodrich Aerospace provided two pneumatic systems, a pneumatic-mechanical de-icer that used rapid bursts of high-pressure air to debond ice and a small tube conventional pneumatic system. Dataproducts New England, Inc., supplied an electro-expulsive system that employed two conductors, overlaying each other in an elastromeric blanket, to generate the repulsive forces necessary to shed ice. The other five systems—developed by Rohr Industries, Advance Concepts De-Icing Company, and Garrett Canada—used electro-impulse (EIDI) coils to shatter and debond ice.

NASA provided a 6-foot span of NACA 0012 airfoil, placed vertically in the IRT at a 4° angle of attack, for the tests. The various systems were subjected to glaze ice conditions at 25°F and rime ice at 10°F through a range of spray times and cycling rates. Researchers measured the power the system required, the residual ice that remained after activation, and the size of the particles that had been shed. The last measurement proved the biggest challenge. The expelled particles could be ingested and damage an engine, so it was critical to know their size. Shin developed a new technique to obtain this vital information, using an Ektopro 1000 high-speed video camera to obtain digitized CRT screen images of the ice particles, together with a motion analysis software package. The instrumentation worked so well that it would be used in all future IRT tests that required this type of information.[26]

The IRT tests yielded a massive amount of useful data on the low-impulse de-icing technologies. With this information in hand, the Air Force and Rockwell International selected two Goodrich systems for further examination in the summer of 1991 on models of inlet lip components of a B1-B engine. Rockwell, the manufacturer of the newest Air Force bomber, wanted to test the Goodrich Pneumatic Impulse Ice Protection (PIIP) system that had performed well in the previous summer's experiments. Goodrich also developed an Electro-Expulsive De-icing System (EEDS), which employed electro-magnetic repulsion to generate the forces necessary to shed ice, for specific use on the B-1B.

Bond subjected the two systems to tests at 230 miles per hour over a range of icing conditions that produced glaze and rime ice. Seven different cycling times, from 12 to 240 seconds, permitted an extensive examination of the effects of the various intervals. Bond was primarily interested in documenting the size of the shed ice participle and the

---

[25] Reinmann, "Cooperative . . . Programs."

[26] Bond and Shin, "Advanced Ice Protection Systems Test in the NASA Lewis Icing Research Tunnel," NASA TM 103757 (1991).

Figure 6–14. John Reinmann in the IRT control room, 1992. (NASA C-92-8515)

distribution, texture, quantity, and thickness of the remaining ice. Data from the tests showed that the PIIP system performed slightly better than the EEDS. However, neither system had been optimized for the complex geometry of the B-1B inlet. Bond concluded that further tests would be necessary to determine the efficiency of the systems.[27]

Bond registered his satisfaction with the results of the two-year program with the Air Force. "The tests were the first of their kind in terms of placing identical constraints on a range of new ice-protection systems," he emphasized. "As a result, an extensive database for residual ice, particle size, and input power and pressure information had been assembled. One of the outcomes of the tests has been the development of a set of test techniques, including a methodology that allows the capture and analytical post-test processing of ice-shedding events."[28]

---

[27] Bond and Shin, "Results of Low Power Deicer Tests on the Swept Inlet Component in the NASA Lewis Icing Research Tunnel," NASA TM 105968 (1993).

[28] Bond, "Low-Power Deicer Program," NASA-Industry Workshop on Aircraft Icing, NASA Lewis Research Center, 27–29 July 1993.

As Reinmann neared retirement in the early 1990s, he had the pleasure of witnessing the IRT become an even more attractive research facility when its ability to simulate high speeds was increased by over 25 percent. This unexpected improvement came in the wake of the "Great Fan Blade Crisis" of the mid-1980s.

On 9 July 1985, the wooden fan blades in NASA's 7 x 10-foot High-Speed Wind Tunnel at Langley suffered a catastrophic failure. To say that this caused great concern at NASA Headquarters would be an understatement. There were ten NASA tunnels that used wooden fan blades, six at Langley, three at Ames, and the IRT. Committees to review the condition of the fan blades were formed at all three centers, together with a national committee on structural integrity of wooden fan blades.

At Lewis, James H. Diedrich was appointed to head the local committee. Other members included Thomas O. Cressman for structural design and analysis; John L. Shannon, Jr. for fracture mechanics and fatigue; Kerry L. Remp for systems safety; and David A. Spera for wooden blade structures. The committee reviewed the history of the IRT blades. In operation since 1944, the fan system had not needed any major components replaced. The fan, 25.4 feet in diameter, consisted of twelve blades of Sitka spruce, each weighing 250 pounds. At maximum revolutions per minute of 460 and a tip speed of 612 feet per second, the fan produced an airspeed of 300 miles per hour without a model in the tunnel. Unlike the fan blades in NASA's other tunnels, the IRT blades had been subject to ice damage. This damage, however, was easily repaired by use of a structural adhesive. The IRT blades also were attached to the hub by bolts and clamping. In the Langley tunnel that experienced the failure, the blades had been secured by bolts alone.

The Lewis review committee reported in December 1986 that the IRT fan blades were not in any danger of failing. The hub attachment system provided "high margins of safety against fatigue failure," and no significant loss of structural integrity could be expected in the next five to ten years of operation at the design speed of 460 revolutions per minute. From the standpoint of fatigue life analysis, the committee concluded, the IRT blades were judged to have "an indefinite life" at 460 revolutions per minute.[29]

NASA Headquarters was not convinced and wanted to shut down the tunnel until metal fan blades were installed. Facility manager Zager countered that the Lewis review had demonstrated that the wooden blades were perfectly safe and argued strongly against a shutdown. Finally, upon the recommendation of the national wood fan blade committee, Headquarters directed Lewis to purchase (for $200,000) and install a new set of blades. After further discussions with the experts at Lewis, however, Headquarters allowed Zager to keep the original blades in operation and store the new blades. At the

---

[29] Diedrich, et al., "Report of the NASA-LeRC Review Committee Regarding the Structural Integrity of the Icing Research Wind Tunnel Fan," 22 December 1986, History Office, GRC.

same time, Headquarters insisted that the fan blades in the IRT be inspected before and after every test.[30]

The inspection dictat, Zager contends, contributed to an accident in the tunnel. Inspection of the fan blades was performed with the aid of a fiberglass ladder. Once, at shift change, the ladder was left in front of the blades and was destroyed when the fan was turned on. Word soon reached Headquarters about a "major accident" in the tunnel. General Billie J. McGarvey, in charge of all facilities, called Zager and wanted to know how long the tunnel would be shut down and how much repairs would cost. Zager told him that the Lewis wood shop would repair the damage to the blades in three or four days—at no cost to Headquarters. McGarvey could not have been more pleased. In the end, it took two weeks for Headquarters to approve the repairs, which were done in four days.[31]

The fan blade study that had been done by the Diedrich committee yielded some unexpected results. Spera, who had assisted in the analysis that had certified the original blades for continued use, had discovered during his calculations that there might be a way to increase the efficiency of the fan. At 460 revolutions per minute, the fan used only 2,800 of the motor's 5,000-horsepower potential to produce a maximum test section velocity of 300 miles per hour. Spera's analysis showed that if the blades were re-machined with a 5° increase in pitch, the same 460 revolutions per minute would increase test section velocity to 385 miles per hour.

In 1993, what Reinmann described as "Spera's great idea" was brought about through in-house work at Lewis. The new fan blades were taken out of storage and modified by the Lewis model development and machining branch. Supports were added to the turning vanes in the diffusing sections of the tunnel. Spera's calculations proved accurate, giving the IRT the higher range of speed that many customers had been requesting.[32]

Reinmann, who in 1988 received NASA's Exceptional Service Medal for his management of the aircraft icing program, retired in May 1994. Reflecting on the major accomplishments of his sixteen years as head of NASA's icing research, he believed that his association with the FAA's Richard Adams on the testing of ground de-icing fluids had contributed significantly to the resolution of an important safety issue. Reinmann also pointed to his support of the flight testing, the modernization of the IRT, the development of cooperative programs with industry, and a grant program to universities that resulted in the training of engineers for icing research, many of whom were subsequently

---

[30] Zager interview.

[31] Ibid.

[32] David Sheldon, "Work Plan: Blade Pitch Modification to the Icing Research Tunnel Drive Fan," 24 February 1993, History Office, GRC.

hired by private companies. In addition, Reinmann had fostered cooperation among researchers, facilitating the exchange of ideas.

One area stood out. During his sixteen years in the icing branch, Reinmann had seen NASA become a key member of an emerging icing community. Together with his deputy, Shaw, he had worked closely with the icing committees in SAE, AIAA, and NATO's Advisory Group for Aerospace Research & Development (AGARD). NASA researchers had served on these committees, presented papers at meetings, and promoted the exchange of information and ideas about developments in icing. Lewis had hosted a number of symposiums and workshops that brought together individuals from private companies and government agencies, both in the United States and abroad, who shared a common interest in icing. Reinmann, of course, had not been alone in these efforts. Adams of the FAA, for example, vigorously had promoted the same objectives. Nonetheless, NASA had been a powerful moving force in creating a close-knit group of the icing experts that formed the heart of the icing community.[33]

Clearly, NASA's icing research program had come a long way since 1978. Compared to the propulsion programs at the Lewis laboratory, it remained modestly funded and manned. "I think management tolerated it," Reinmann recalled, "because it produced the type of publicity that the public and Headquarters could understand—flight safety. And the IRT's uniqueness and its heavy usage by industry added to [Lewis's] value when Congress was looking to eliminate some of the NASA centers." Indeed, by the time of Reinmann's retirement, the IRT was being used more than a thousand hours a year, on average. "That high usage rate," Reinmann emphasized, "says a lot for our tunnel operators, our operations engineers, and our research engineers." Important work had been done in developing advanced ice-protection systems and in formulating computer codes to predict the response of any aircraft to an icing encounter. Much had been accomplished under Reinmann's leadership, but new challenges lay ahead.[34]

---

[33] Reinmann interview.

[34] Interview with Reinmann in *Lewis News*, 26 March 1993; Reinmann to Leary, 12 April 2001.

Chapter 7
New Challenges

While the major icing safety issue of the 1980s had centered on the ground de-icing problems of large transports, the 1990s saw the focus shift to inflight icing difficulties that were encountered by the smaller airplanes of the booming commuter aviation industry. The Airline Deregulation Act of 1978 had provided the legal foundations for a sharp rise in short-haul traffic between large metropolitan areas. This had been followed by "code-sharing" agreements in the mid-1980s, under which commuter carriers acted as "feeders" for major airlines. As a result, passengers carried by the commuters had risen from 11 million in 1978 to 38 million in 1990.[1]

American Eagle, a feeder for American Airlines, had emerged during the 1980s as one of the larger commuters. Starting out in 1979 as Simmons Airlines, a small commuter based in Marquette, Michigan, the company had concluded a code-sharing agreement with American Airlines in April 1986 and had begun flying as American Eagle. Two years later, Simmons was purchased by the parent company of American Airlines. By the early 1990s, American Eagle was a major airline in its own right, operating a fleet of seventy-nine aircraft and flying to thirty cities from American Airlines's Chicago hub and thirty-one cities from the parent's Dallas-Fort Worth hub.[2]

Prominent in American Eagle's fleet were forty-seven aircraft that had been built specifically for the commuter market by Avions de Transport Regional (ATR). Formed in 1980 by Aerospatiale of France and Aeritalia of Italy to jointly develop regional airliners, ATR had first produced the model 42 in 1984. The ATR 42 was a twin-engine turboprop, with a T-tail and high straight wing that carried forty-six passengers and crew.

---

[1] On the rise of the commuter industry, see R. E. G. Davies and I. E. Quastler, *Commuter Airlines of the United States* (Smithsonian Institution Press, 1995).

[2] "Aircraft Accident Report: In-Flight Icing Encounter and Loss of Control, Simmons Airlines, d.b.a. American Eagle Flight 4184, Avions de Transport Regional (ATR) Model 72-212, 31 October 1996," NTSB/AAR-96/01, adopted 9 July 1996.

Figure 7–1. William Sexton de-ices model in the IRT, 1992. (NASA C-1992-10571)

This had been followed by the ATR 72, a stretched version, 14 feet, 8 inches longer than the original model, that could accommodate 72 passengers and crew. American Eagle operated twenty-five ATR 42s and twenty-two ATR 72s.

The ATR 42/72 was certified for flight into forecast icing conditions, having demonstrated the ability to operate safely in the maximum continuous and maximum intermittent icing envelopes specified in Part 25, Appendix C of FAA Regulations. To protect against icing, the ATR-42/72 had pneumatic boots on the leading edges of the wing, as well as the horizontal and vertical stabilizer. A pneumatic system also protected the engine air inlets, while electric heating was used for propeller blades, windshield, pitot tubes, and balance horns of the ailerons, elevator, and rudder. The airplane was equipped with an Ice Evidence Probe, outside and below the left cockpit window, which gave the pilot visible evidence of ice accretion. In addition, there was a state-of-the-art Rosemount ultrasonic ice-detector probe mounted on the underside of the leading edge of left wing, between the pneumatic boots, that gave both a visual and aural alarm of ice accretion.

The ATR 42/72 ice-protection system was designed to be operated at three levels. Level I, which activated all probes and the windshield heating system, was turned on at all times. When a pilot encountered atmospheric conditions that were conducive to icing, he went to Level II, which turned on the engine intake boots and the electrical system for propellers and aileron/elevator/rudder horns. At the first visual indication of icing, the pilot went to Level III and activated the pneumatic boots for the wings and horizontal and vertical stabilizers.

On 31 October 1994, American Eagle Flight 4184 departed Chicago's O'Hare Airport on schedule at 11:39 A.M. for a round-trip flight to Indianapolis. The ATR 72 reached its destination without incident. The flight left Indianapolis at 2:55 P.M. for the 45-minute return to Chicago—the pilot having been informed that icing might be encountered en route. At 3:13 P.M., Captain Orlando Aguiar received clearance for his approach into Chicago and began to descend from 16,000 feet to 10,000 feet. During the descent, he activated the Level III ice-protection system.

Upon reaching 10,000 feet, Chicago Approach Control directed flight 4184 to enter a holding pattern, 18 miles south of the Chicago Heights VOR. As the airplane went around in its racetrack holding pattern, everything seemed normal. A flight attendant visited the cockpit and discussed "both flight and non-flight-related information with the pilots." At 3:30 P.M., Captain Aguiar remarked to first officer Jeffrey Gagliano, "we're just wallowing in the air right now." Three minutes later, Aguiar set the flaps to the 15° position, but icing did not seem to be a major concern. "That's much nicer, flaps fifteen," Gagliano commented to Aguiar. At 3:49 P.M., Aguiar departed the cockpit and went aft to use the restroom. When he returned, Gagliano remarked, "we still got ice."

At 3:56 P.M., flight 4184 was cleared to descend to 8,000 feet and to expect clearance into Chicago in 10 minutes. As the aircraft approached 9,000 feet, it commenced to roll rapidly to the right, and the autopilot disconnected. The cockpit voice recorder picked up "brief expletive remarks," followed by "intermittent heavy breathing." At 77° right wing down, the ATR 72 started to roll left toward wings level, then right again at a rate in excess of 50° per second. Captain Aguiar's control column force exceeded 22 pounds as the airplane made a complete roll through an inverted position to wings level. The ATR 72 then began another roll excursion to the left. At this point, the airspeed registered 300 knots with 73° nose down, as both pilots pulled on the control column.

As the airplane reached 1,700 feet, the "Whoop! Whoop!" of the ground proximity warning system sounded. First officer Gagliano made "an expletive comment" as the airspeed increased to 375 knots, 38° nose down. The crew continued to fight for control of the aircraft until the ATR 72 crashed into a wet soybean field near Roselawn, Indiana, at 3:59 P.M. The two pilots, two flight attendants, and all 64 passengers died instantly.

On 8 November 1994, the National Transportation Safety Board (NTSB) contacted the icing branch at Lewis and asked for technical support in its investigation of the Roselawn tragedy. The NTSB was especially interested in obtaining information on ice accretion, ice-protection systems, and data on iced airplane stability and control. NASA responded to assigning researchers Jaiwon Shin and Thomas P. Ratvasky to the NTSB's Airplane Performance Group.[3]

Shin and Ratvasky traveled with the group to Toulouse, France, on 12 November. For a week, the group met with the ATR engineering staff to review icing certification documents and test procedures, and to discuss previous roll-control anomalies due to icing. Shortly after Shin and Ratvasky returned to the United States in early December, the NTSB asked NASA to conduct tests in the IRT of an MS-317 airfoil, which had a similar cross-section profile to the outboard portion of an ATR 72 wing. The tests, conducted on 8–10 and 12–14 December, showed that large super-cooled water droplets, which had been detected in the vicinity of the Roselawn accident, could accrete beyond the area that was protected by pneumatic boots. Even before the tests were completed, the FAA had sufficient concern about the problem to prohibit ATR 42/72s from flying in "known or forecast" icing conditions.

While NASA developed a program to look more carefully in the phenomenon of large droplet icing, an FAA special certification review team arranged with the U.S. Air Force to use their KC-135 spray tanker for flight tests with an ATR 72. Again, the Board requested the technical assistance of the icing section. This time, Richard Ranaudo, an experienced Twin Otter research pilot, accompanied Ratvasky to Edwards Air Force Base, California, the site of the tests.

The tanker flights began on 14 December. Ranaudo's experience soon proved valuable to the test team. For example, he suggested using tufts on the upper surface of the ATR 72 wing to help detect flow separation. At the same time, Ratvasky held discussions with Air Force personnel about icing cloud characteristics and measuring techniques.

The tanker tests demonstrated that in large droplet conditions with flaps at 15°, a ridge of ice could form at the 9 percent chord position of the ATR 72's wings, whereas the aircraft's pneumatic boots covered only to the 7 percent chord. This problem no doubt caused the uncommanded roll event at Roselawn. On 11 January 1995, the FAA withdrew its blanket prohibition against flight into icing conditions for the ATR 42/72, pending the fitting of new pneumatic boots to cover a larger wing area and new training for pilots.

The Roselawn investigation marked the beginning of a major program at Lewis to study the icing conditions found in freezing drizzle and freezing rain, collectively known

---

[3] Icing Technology Branch to the director of aeronautics, "Summary of NASA LeRC Involvement in the ATR-72 Accident in Roselawn, IN," 5 January 1995. The author is indebted to Mr. Ratvasky for a copy of this document.

as super-cooled large droplet (SLD) icing. Up to this point, the goal of the IRT researchers had been to simulate the natural icing conditions that had been defined by the flight research of the late 1940s. These conditions, as noted earlier, had been incorporated into Appendix C of FAR Part 25 and set the parameters that aircraft had to deal with to be certified for flight into forecast icing conditions. One component of the icing environment was droplet size. The great majority of icing involved drops with a mean effective diameter of 15 to 50 microns. The early challenge for the nozzle designers of the IRT had been to produce these small diameter droplets.[4]

Icing conditions, of course, had always existed in nature beyond the Appendix C envelope. The FAA, however, assumed that the probability of aircraft flying into these extreme conditions was one in a thousand. Even before the Roselawn accident, questions had been raised about the continued acceptability of the Appendix C standards. The appearance of a new generation of turboprop airliners in the 1980s made these questions more urgent, as their advanced, highly efficient wings were less tolerant of icing than previous airfoils. Even a small amount of ice could make these turboprops difficult to control—without the warning signs that pilots had come to expect.[5]

NTSB investigators estimated that the American Eagle ATR 72 had flown into super-cooled water droplets in the size range of 100 to 400 microns. These conditions, FAA safety engineer John P. Dow, Sr., pointed out, "may challenge contemporary understanding of the hazards of icing." While the FAA launched its own program to study the safety and certification implications of SLD icing, NASA researchers Harold E. Addy, Jr., and Dean R. Miller, together with Robert F. Ide of the Army Research Laboratories, began a series of experiments in the IRT to determine the effect of SLD ice accretion on Twin Otter and NACA 23012 wing sections.[6]

---

[4] Whereas Appendix C used "mean effective droplet diameter" (MED) as a standard of measurement, researchers in the 1990s with their superior measurement tools used "median volumetric diameter" (MVD). Both MED and MVD divided total water volume present in the droplet distribution in half, with half the water volume in larger drops and half in smaller drops, but MED was based on assumed droplet distribution while MVD measured actual droplet size.

[5] For the problems facing pilots, see Jan W. Steenblik, "Inflight Icing: Certification vs. Reality . . . Where the Difference Can Mean Life or Death." This article first appeared in *Air Line Pilot* in August 1995 and was reprinted in Flight Safety Foundation, "Protection Against Icing: A Comprehensive Overview," a special issue of *Flight Safety Digest* (June–September 1997), pp. 154–60.

[6] For Dow's remarks and the FAA's program, see "Proceedings of the FAA International Conference on Aircraft Inflight Icing," DOT/FAA/AR-96/81,l (August 1996), vol. I, pp. 7–11. NASA's program is detailed in HaeOk Skarda Lee, "NASA's Aircraft Icing Research in Supercooled Large Droplet Conditions," in ibid., pp. 83–100.

Figure 7–2. Technician Dan Gideon with AH-64 (Apache) Longbow fire control radar during icing tests, 1993. (NASA C-1993-4831)

Using a droplet size of 160 microns, Ide, who was responsible for calibrating the tunnel's icing cloud for droplet size, liquid water content, and cloud uniformity, ran a special series of calibration tests. Although he did not have time to calibrate the 160-micron-diameter cloud at various liquid water contents for given airspeeds, he was satisfied that the tunnel simulation would produce accurate results.[7]

The first series of experiments employed a Twin Otter wing section. Researchers subjected the aluminum airfoil with a constant chord of 77.25 inches to an array of large droplet icing conditions, including variations in temperature, airspeed, droplet size, angle of attack, pneumatic boot cycle interval, and flap settings. They found that an ice ridge formed aft of the de-icer boot on every experimental icing run. The ridge was sensitive to changes in temperature, reaching a maximum size at 28°F for 125 miles per hour and 30°F for 195 miles per hour. Variations in boot cycling time did not have a significant effect on the residual ice accretion.

A second series of IRT tests were performed on a NACA 23012 wing section. This tapered airfoil was mounted vertically in the test section and had a chord length which varied from 73.8 inches at the floor to 65.2 inches at the ceiling. Technicians outfitted the leading edge of the model with a full-span pneumatic de-icer boot that extended to 6 percent of the chord on the suction surface and to 11 percent on the pressure surface of the airfoil.

Researchers Addy and Miller selected the following "anchor points" for the tests: 32°F, 160 microns, 0.82 grams per cubic meter LWC, 195 miles per hour, 0° angle of attack, 3 minutes boot cycle, and 18 minutes spraying time. They then investigated the effects of each one of the anchor points while holding the other points steady. For example, what would be the effect on the airfoil of lowering the temperature to 28°F while all other conditions remained unchanged for the anchor points?

As in the previous Twin Otter study, researchers discovered that temperature had the most significant impact on ice accretion. While holding the other parameters constant at their anchor points, the researchers ran tests at temperature ranges of 5°F to 37°F. At 37°F, all impinging water ran back to the trailing edge and was blown off the airfoil. Ice began to form at 34°F, with a distinct ridge just aft of the active portion of the boots on both suction and pressure surfaces. At temperatures below 30°F, the ridge appeared to harden and become more resistant to shedding. The largest ice ridge—over 1 inch in height—occurred in temperature ranges of 24°F through 28°F.

Tests on the Twin Otter and NACA 23012 airfoils, the researchers concluded, yielded "valuable information, as well as a substantial database on large droplet ice accre-

---

[7] Harold E. Addy, Jr., Dean R. Miller, and Robert F. Ide, "A Study of Large Droplet Ice Accretion in the NASA Lewis IRT at Near Freezing Conditions; Part 2," NASA TM 107424 (1996).

tions." Nonetheless, there remained a great deal of work to be done. The next phase of research into the phenomenon of SLD icing, they believed, should be "a fairly rigorous flight program." Because so little data was available to characterize naturally occurring large droplet icing conditions, flight tests were necessary "to generate a realistic large droplet cloud envelope for the IRT and to help define tests to investigate the effects of varying LWC on large droplet ice accretions."

Following the IRT tests, NASA and the FAA agreed to co-sponsor a three-year program of flight research to develop a database to document atmospheric large droplet icing, improve weather forecasting tools, extend icing simulation capabilities of the IRT and LEWICE, and provide educational information to pilots to help them deal with large droplet icing encounters. Joining NASA and the FAA in this program was the National Center for Atmospheric Research (NCAR), which would provide weather forecasts and inflight guidance. In addition, Atmospheric Environmental Services, Robotic Vision Systems Incorporated, and BFGoodrich Aerospace provided special equipment for NASA's Twin Otter.[8]

Target areas for the research program would be the Great Lakes region. Not only did this location have conditions that were conducive to large droplet icing, but it also was seeing an increasing number of commuter flights. NCAR scientists examined meteorological data for the Cleveland area for the years 1961 to 1990 and concluded that the highest probability for encountering SLD conditions occurred during the months of December through March.

NASA conducted research flights in 1997 between 13 January and 5 February, and between 6 March and 3 April. In all, twenty-nine flights took place. Four flights were conducted in clear air in order to establish an ice-free baseline. Most of the flights into icing encountered a mixture of small and large droplets. Three flights, however, were made in prolonged large droplet conditions.

A typical mission for this program began with preflight checks of the aircraft and instrumentation at 6 A.M. This was followed by a weather briefing from NCAR at 7 A.M. Forecasters used a variety of information in an attempt to locate icing areas. Infrared satellite imagery determined cloud locations, cloud top temperature, cloud top altitude, and the likelihood of liquid water versus ice crystals, while radar identified areas of precipitation. Forecasters also had surface observations, surface and upper air charts, and pilot reports. They integrated this information to identify large droplet and other icing areas, then suggested a flight location, strategy for sampling the clouds, and an escape

---

[8] Dean Miller, Thomas Ratvasky, Ben Bernstein, Frank McDonough, and J. Walter Strapp, "NASA/FAA/NCAR Supercooled Large Droplet Icing Flight Research: Summary of Winter 96–97 Flight Operations," NASA TM 206620 (1998).

route. Researchers were especially impressed with the forecasting talents of meteorologist Ben C. Bernstein. "Ben's knowledge and skill," Ratvasky has emphasized, "were the main reasons we were able to get into SLD." After discussing the proposed flight plan with NCAR, the flight crew would prepare for a takeoff between 8 and 9 A.M.

Climbing out of Cleveland, the Twin Otter would head toward the target area at an altitude prescribed in the weather briefing. While en route, the crew would be in contact with NCAR meteorologists via a satellite communications link to discuss the most recent information. When the aircraft reached the area of forecast large droplet icing, the crew would attempt to perform a "sounding" of the cloud. They wanted to know the cloud top and base, as well as the altitudes of the freezing level, liquid water, and large droplet icing.

With the sounding completed, the aircraft entered the icing cloud at which it was most likely to find SLD conditions. Researchers would then sample the cloud during a series of horizontal transects. At some point, the crew would decide to exit the cloud to document icing accretion with a stereo photographic system and to test its effect on aircraft performance.

At the conclusion of the flight, a debriefing session would be held with NASA's pilots, operations staff, researchers, and NCAR meteorologists. Participants would review how the flight had been conducted, identify areas for improvement, and discuss instrumentation problems. Researcher Dean Miller would then analyze the cloud physics data—droplet spectra, liquid water content, temperature—and send the results to the FAA's Technical Center, where Richard K. Jeck was reviewing Appendix C criteria in order to expand the icing certification envelope to include SLD conditions.

The second phase of the flight research program, which took place between November 1997 and March 1998, turned out to be the most successful icing season of the three-year program. Researchers made a total of fifty-four flights, encountering icing conditions during twenty-eight of eighty-three flight hours. The high point of the program occurred on 4 February 1998, when the Twin Otter experienced a prolonged exposure to freezing rain. Whereas most SLD icing had focused on freezing drizzle, little was known about the effects of larger droplet freezing rain on aircraft performance. In one previous encounter with the phenomena by a research aircraft, ice accretion had adhered smoothly to the airfoil surface and had not caused any significant degradation of performance. It was a far different story, however, for the SLD researchers.

While flying in the vicinity of Parkersburg, West Virginia, the Twin Otter had a 90-minute exposure to freezing rain, during which droplet sizes exceeded 1,000 microns. Even with de-icing equipment turned to "auto-fast," in which the boots cycled every 60 seconds, clear ice remained on and aft of the boots, while ice ridges built up on the wings and tail. Climbing out of the cloud deck, the Twin Otter conducted three performance sweeps. They showed that maximum airspeed had declined from a clear airspeed of 150 knots to 115 knots, while buffet speed went from 68 knots to 90 knots. While the flight envelope

Figure 7–3. IRT renovation, 1993. (NASA C-93-7170)

decreased by 70 percent, coefficients of drag increased by 60 to 200 percent (depending on angle of attack), and maximum coefficient of lift was reduced by 30 percent.

"This case study," the report of the flight emphasized, "clearly demonstrates that exposure to classical FZRA [freezing rain] can cause significant ice to accrete on the Twin Otter airframe, including ice ridges and nodules on unprotected surfaces." Subsequent performance tests revealed "a dramatic increase in aircraft drag and decrease in lift that decreased the size of the Twin Otter's safe operating window." Freezing rain, the researchers concluded, "can pose a significant flight icing hazard and should not be ignored when considering SLD issues."[9]

The third year of flight research proved anticlimactic. The program started late, in February 1999, due to a Headquarters-mandated safety inspection in January, and it lasted only until March. A total of twelve flights were conducted with 15 flight hours. Instrument problems plagued most missions, and little useful data was obtained. Whereas Miller had been able to send 828 data points to Jeck during the previous two phases of the SLD flight program, he was unable to pass along any useful data from the final effort.[10]

Overall, the SLD flight program was an outstanding success. Thanks to the efforts of researchers, pilots, NCAR meteorologists, electrical engineers, avionics technicians, and fabrication, imaging, and data-processing specialists, a great deal had been learned about SLD icing conditions. At the FAA's Technical Center, Jeck took the data from the NASA flights, along with information from the Meteorological Service of Canada and other sources, to produce a combined set of SLD data for review by the Ice Protection Harmonization Working Group (IPHWG), an international body that was chartered by the FAA's Aviation Rulemaking Advisory Committee. "The IPHWG is still working with the issue of SLD icing conditions in the certification process," Ratvasky noted in May 2001, "but the data that has been provided by NASA and MSC has enabled the policy-makers to make decisions based on solid research."[11]

Tailplane icing ranked with large droplet icing as the most notable icing safety issue of the 1990s. The problem, itself, like SLD icing, was not entirely new. For example, undetected ice on the horizontal stabilizer of a Continental Airlines Vickers Viscount 810 had caused the airplane to pitch down during an attempted night landing in freezing precipitation at Kansas City on 29 January 1963. The aircraft had struck the ground short of the runway, causing eight fatalities. The problem, however, became more urgent

---

[9] Ben C. Bernstein, Thomas P. Ratvasky, Dean R. Miller, and Frank McDonough, "Freezing Rain as an In-Flight Icing Hazard," NASA TM-2000-210058 (June 2000).

[10] Ratvasky to Leary, 11 May 2001.

[11] Ibid.

with the appearance of the new generation of turboprop commuter airplanes in the 1980s. These aircraft had advanced airfoils that were unforgiving when it came to ice contamination and featured non-powered reversible control surfaces and pneumatic boot de-icing systems. With their short-haul routes that were generally flown at lower altitudes, the commuters had a greater exposure to icing than long-haul jet transports. This combination of factors made them vulnerable to tailplane icing.[12]

Research into the problem had first been undertaken during the late 1970s and early 1980s by a Swedish-Soviet working group in the wake of the crash to a Linjeflyg Vickers Viscount on 15 January 1977, near Stockholm. Their wind tunnel and flight tests that studied the tailplane stall phenomenon had failed to produce definite results, but had provided insights that would assist later researchers. In the early 1990s, NASA had conducted a preliminary flight investigation into control anomalies that were caused by tailplane contamination. Researchers found that an ice casting placed on the Twin Otter's horizontal tail had caused decreased longitudinal stability, a condition that had worsened when flaps had been lowered 10° or more.[13]

To promote greater awareness of the problem, NASA and the FAA sponsored an International Tailplane Icing Workshop at Lewis in early November 1991 that was attended by more than 100 representatives of industry and government from the United States and eight foreign countries. It soon became clear that the phenomenon of ice accretion on the horizontal tailplane of turboprop commuters was widespread but not well understood, especially in the United States, by either manufacturers or operators. The Europeans seemed somewhat better acquainted with the problem. The European Joint Airworthiness Authority, for example, required manufacturers to demonstrate a specific pushover maneuver during certification for tailplane icing, whereas the FAA did not.

Captain Stephen Ormsbee of Henson Airlines, chairman of the Air Line Pilots Association's Regional Airline Committee, pointed out that the chances of a turboprop flight crew experiencing tailplane icing were far greater than their chances of encountering severe wind shear. "We need to educate pilots about the hazards associated with tailplane icing and provide them with a reliable means of detecting its presence," he noted in the final report of the workshop. Also, he continued, "we need to require that turboprop airplanes demonstrate their ability to handle tailplane icing during certification."[14]

---

[12] Porter J. Perkins and William J. Rieke, "Aircraft Icing Problems—After 50 Years," a paper prepared for the 31st Aerospace Sciences Meeting, Reno, NV, 11–14 January 1993; copy courtesy of Mr. Perkins.

[13] T. P. Ratvasky and R. J. Ranaudo, "Icing Effects on Aircraft Stability and Control Determined from Flight Data, Preliminary Results," NASA TM 105977 (1993).

[14] Ormsbee is quoted in Jan W. Steenblik, "Turboprop Tailplane Icing," *Air Line Pilot*, January 1992, pp. 30–33.

Following a second workshop in April 1993, the FAA made a formal approach to NASA for assistance "to conduct research into the characteristics of ice-contaminated tailplane stall (ICTS) and develop techniques or methodologies to minimize this hazard." Tailplane icing, Thomas E. McSweeney, director of the FAA's Aircraft Certification Services, wrote to NASA, was "strongly suspected" as the cause of sixteen accidents during the past years that resulted in 139 fatalities. Most of these accidents had involved turbo-prop-powered transport and commuter aircraft, of which there were now 1,500 operating in the United States. Because of the implications for flight safety, McSweeney proposed that the FAA and NASA undertake a joint program to assure that the problems associated with tailplane icing were identified and corrected before an airplane entered service.[15]

The FAA's request led to the signing of an interagency agreement under which NASA's Lewis Research Center and the FAA Technical Center at Atlantic City co-sponsored a Tailplane Icing Program (TIP). Thomas Ratvasky was given the task of drawing up a work plan for the program. He would prove an excellent choice for the assignment. A graduate of Case Western Reserve University in 1988, he had joined NASA's icing branch in 1990 after receiving his master's degree in aeronautics from George Washington University. His first project had been a stability and control program to observe the effects of artificial ice shapes on the tail of the Twin Otter. Later research had involved flights into natural icing conditions. The TIP work would be a natural continuation of these earlier research efforts.[16]

Early in 1994, representatives from Lewis, the FAA Technical Center, FAA Certification Services, and Ohio State University met to discuss the work plan that Ratvasky had developed. They confirmed the NASA/FAA Tailplane Icing Program which would be funded by both agencies. Also, they endorsed Ratvasky's proposal to enter into a cooperative agreement between NASA and Ohio State University for dry air testing. Finally, they concurred on a four-year program that would involve both icing and dry air tunnel tests, followed by flight tests. Test facilities included the IRT, Ohio State's Low-Speed Wind Tunnel, and NASA's Twin Otter. Ratvasky would be the lead NASA researcher. Richard Ranaudo, a former U.S. Air Force fighter pilot who had joined NASA in 1974 and had flown in the subsonic and supersonic engine test programs as well as in icing research, was designated as the lead pilot for the flight program. James T. Riley of the FAA Technical Center would monitor the program and lend his expertise where needed.[17]

---

[15] McSweeney to Wesley L. Harris, 21 April 1994; copy courtesy of Mr. Ratvasky.

[16] Interview with Thomas P. Ratvasky by William M. Leary, 22 September 2000.

[17] Ratvasky, "NASA/FAA Tailplane Icing Program Work Plan," 8 April 1994, revised 1 May 1995; copy courtesy of Mr. Ratvasky. Ratvasky presented this plan at the third FAA-sponsored International Tailplane Icing Workshop, held in Toulouse, France, September 1994.

Figure 7–4. IRT fan, 1993. (NASA C-93-7163)

The first step in Ratvasky's work program involved the selection of an appropriate model to test in the two tunnels. The model chosen for the IRT's 6 x 9-foot test section was an actual portion of a Twin Otter tailplane. The flight hardware was cut to provide a 6-foot span that was mounted vertically and attached to the IRT's force-balance system on the turntable and tunnel ceiling. The turntable enabled researchers to change the angle of attack of the vertically mounted airfoil. BFGoodrich Aerospace supplied a new model pneumatic boot, which was installed on the leading edge of the model to determine ice accretion between cycles of the boot. Load cells on the elevator hinge brackets of the airfoil measured elevator hinge movement during the tests.[18]

---

[18] Ratvasky, Judith Foss Van Zante, and James T. Riley, "NASA/FAA Tailplane Icing Program Overview," NASA TM 208901 (1999).

A different model was developed for Ohio State's 7 x 10-foot tunnel. Model makers used machinable plastic (Ren Shape) to construct a 2-D airfoil with a 4.75-foot chord and 7-foot span. It was intended to replicate the two-section geometry of the Twin Otter and included an elevator that could be set at various angles.

Ratvasky next turned to selecting accurate ice shapes for testing. He wanted to begin in the IRT, but the NASA facility was so heavily scheduled that he decided to use Ohio State's tunnel for the initial aero-performance tests. One ice shape to be attached to the model had been developed from one of NASA's icing flights in 1984. Another was derived from LEWICE predictions. The two shapes were machined from Ren Shape, then sanded smooth and painted.

Initial tests at Ohio State began in September 1994. Data on the model was first taken without ice shapes in order to establish a baseline. The two ice shapes were then attached to the airfoil, and measurements were made at various airspeeds, elevator angles, and angles of attack. Just as the Ohio State tests concluded, the American Eagle accident at Roselawn occurred, focusing attention on the more urgent problem of large droplet icing. As noted earlier, Ratvasky joined the NTSB team that investigated the ATR 72 accident.

Ratvasky was not able to turn his attention to the Tailplane Icing Program until the fall of 1995. The next phase of the program involved flight tests to establish a baseline against which to measure the effects of tailplane contamination. In September and October, the Twin Otter made seventeen flights. While pilots Ranaudo and Dennis O'Donoghue confirmed the operation of the airplane's hydraulic system during low-gravity maneuvers, Ratvasky measured tail-flow conditions and elevator settings for cruise, hold, and approach, as these conditions would be simulated in impending IRT tests. They determined the range of tailplane inflow angles for the flap, speed, thrust, and normal accelerations to be tested with ice shapes.

Ratvasky was able to conduct TIP tests in the IRT in February 1996. The Twin Otter airfoil was subjected to a variety of icing cloud conditions, while different airspeeds, angles of attack, and elevator deflections simulated the cruise, holding, and approach phases of flight. Following each test run, technicians made ice tracings and took photographs of the resultant ice shapes. At the end of the night, they made a mold of a 15-inch span of the final ice accretion. Skilled craftsmen in the model shop then extracted multiple polyurethane castings from the mold to have a full-span ice accretion for later aero-performance tests in the Ohio State tunnel in August 1996.

Finally, everything was ready for flight tests with the various ice shapes that had been developed in the IRT and Ohio State tunnel. The climax of the test program came between July and September 1997. Researchers Ratvasky and Judith Foss Van Zante took data as pilots Ranaudo, Robert McKnight, and William Rieke searched for the boundaries of tailplane stalls with various ice shapes. On 17 September, Ranaudo passed the boundary. The Twin Otter experienced a full tail stall, pitched nose down, and lost 300 feet before Ranaudo recovered control, despite immediate corrective action.

A dozen flight tests with three ice shapes revealed that the paths to tailplane stall increased with ice shape severity, flap deflection, airspeed, and thrust. Tactile clues that preceded stall events were difficulty (or inability) to trim the airplane, pitch excursions, and a buffeting of the controls. To recover from the stall, the pilot had to reduce power, pull back on the yoke, and raise flaps.

Tail stall recovery procedure, researchers emphasized, was the opposite of a recovery from a wing stall. Ever since the early days of flight, pilots had been taught that the proper way to recover from a stall was to push down on the yoke and add power. But tailplane stall was different. In a wing stall, airflow separates from the upper surface of the airfoil; whereas, with a tailplane stall, the flow separates from the lower surface of the tail. "Because of these differences in the stalling mechanism and recovery procedures," Ratvasky and his colleagues concluded, "it was determined that pilots should be made aware of the clues that may occur prior to a tailplane stall."[19]

NASA quickly set out to spread the word. In October 1997, the icing branch conducted a guest pilot workshop for fifteen pilots and engineers from the FAA, Transport Canada, Bombardier, Cessna, Raytheon, *Airline Pilot, Aviation Week & Space Technology, Commercial & Business Aviation, Flying Magazine,* and *Professional Pilot.* Each pilot had the opportunity to fly the Twin Otter with a tailplane ice casting, together with a NASA safety pilot. The guest pilots experienced the handling qualities of an aircraft with an ice-contaminated tailplane and practiced the pushover maneuver used in icing certification flight tests.

"The guest pilot program," Ratvasky emphasized, "is important for building good relationships with aircraft manufacturers that in turn allow NASA to provide them with appropriate design tools. The program enables us to get the word out that NASA Lewis is performing relevant aviation safety programs that will impact the whole industry, ranging from the aircraft manufacturers that make the airplanes to the operators who fly them."[20]

The following year, the icing branch produced a 23-minute video on tailplane icing that described why and how tailplane icing occurred, the warning signs of an imminent stall, and the proper recovery procedures. The dramatic highlight of the video came with cockpit film of Ranaudo's full tailplane stall of September 1997. Widely distributed, the video received such an enthusiastic reception from the aviation community that NASA

[19] Ratvasky, Van Zante, and Alex Sim, "NASA/FAA Tailplane Icing Program: Flight Test Report," NASA TP 209908.

[20] *Lewis News,* December 1997.

put out a second video in 1999 that reviewed recent research into icing problems for regional and corporate pilots.[21]

Thomas Bond, who became head of the icing branch in 1998, believes that the videos have made a major contribution in the area of education. They also have been cost-effective. Whereas the Tailplane Icing Program ran for four years and cost about $1.2 to $1.4 million, the video that used the results of the program took about six months to produce and cost about $45,000. For this small budget, NASA received a high degree of favorable publicity, as well as materially aiding aircraft safety.[22]

Ratvasky considered the Tailplane Icing Program to be a perfect example of how the various parts of NASA's icing research interconnected to produce results. The IRT, LEWICE computer codes, and flight research combined to greatly improve the aviation community's understanding of the phenomenon of tailplane icing, its effect on aero-performance, and what could be done to combat it. Flight research, he emphasized, was a crucial component of this effort, as it provided a "truth statement" for simulations.[23]

It came, therefore, as a bitter disappointment when NASA Headquarters decided in 1999 to terminate the flight research program in the interests of cost-cutting. The decision provoked both private and public protests. Icing chief Bond told his superiors at the recently renamed John F. Glenn Research Center that terminating flight operations would lead to "the loss of a world-class facility and the flight research staff core competency that is unparalleled in any organization in the world." Glenn acting deputy director Gerald Barna argued in a memo to NASA Headquarters that the Twin Otter was "research essential" and should remain in service for at least two more years.[24]

David M. North, editor-in-chief of the influential journal *Aviation Week & Space Technology*, spoke for many in the aviation community in a forthright editorial that called into question NASA's priorities. The Glenn Center, he pointed out, had "one of the best refrigerated wind tunnels in the business, and its researchers are the acknowledged leaders in the study of icing on the stability, control, and performance of aircraft." North was able to attest to "the quality and professionalism" of the icing researchers at Glenn, having participated in the guest pilot workshop. His flight in the Twin Otter had been "an eye-opener for the recommended recovery procedures when experiencing tailplane icing."

Icing, he emphasized, had been killing people for more than seventy years and continued to do so. This was not the time to let up on icing research and training. For years,

[21] NASA Lewis icing branch, "Tailplane Icing," 5 September 1998; NASA Glenn icing branch, "Icing for Regional and Corporate Pilots," 20 October 1999.

[22] Interview with Thomas H. Bond by William M. Leary, 22 September 2000.

[23] Ratvasky interview.

[24] *Cleveland Plain Dealer*, 11 January 2000.

NASA had shortchanged research and development in aeronautics in favor of the space program. "While the International Space Station and Mars research may be glitzy and glamorous," he wrote, "NASA should not abandon an established effort that has saved lives and promises to save many more." If NASA could spend billions of dollars on the International Space Station, then surely the agency could spare the $300,000 that it hoped to save by terminating the flight research program.[25]

In January 2000, NASA's Office of Inspector General announced that it would review the decision to end flight research. The ultimate fate of the program remained to be seen as NASA's icing research faced the challenges of a new millennium.[26]

---

[25] David M. North, "Reinstate NASA Icing Research," *Aviation Week & Space Technology,* 152 (31 January 2000): 68.

[26] *Cleveland Plain Dealer,* 29 January 2000.

Chapter 8
The 1990s

The icing branch underwent a number of senior personnel changes during the 1990s. Following Reinmann's retirement in 1994, there were a series of acting branch chiefs for some nine months before Hao Ok Lee took over the job in 1995. Two years later, after Lee moved to a more senior position in the Lewis Base Program Office, Thomas Bond became acting chief, then permanent head of the branch in 1998. A one-time airplane mechanic on Twin Otters in Alaska, Bond received his B.S. in mechanical engineering from Penn State in 1983. He joined NASA upon graduation and started work in the internal combustion engine section. In 1986 Bond moved to the icing branch, where he gained a reputation as a "hands-on" individual who enjoyed experimental work.[1]

IRT facility managers also changed, beginning with Harold Zager's retirement in 1988. Several months before Zager left NASA, David W. Vincent was brought in to relieve him of responsibility for the IRT. A graduate of the University of West Virginia in 1963 with a degree in aeronautical engineering, Vincent had worked in a variety of assignments during his twenty-five-year career at NASA-Lewis, but he had limited tunnel management experience. "Not knowing anything about Dave," Reinmann recalled, "we were a bit concerned when he was assigned to the IRT." Reinmann's apprehension soon vanished, however. Although just as committed as Zager to maximum productivity for the tunnel, Vincent's style of management was less confrontational than his predecessors. With respect to scheduling, for example, Vincent adopted a cooperative approach, with weekly meetings to review and adjust time allocations. Also, he had not been instructed, as had Zager, to make contractors "happy." On the contrary, there was a new emphasis on having manufacturers pay for tunnel time except when they agreed to share data or participate in a joint program with NASA researchers. Vincent would not allow contractors to have additional time in the tunnel to complete experiments, although he would expand daily tunnel oper-

---

[1] Bond interview; Reinmann interview.

Figure 8–1. Icing Branch in 1993. Standing (L/R): Mario Vargas, John Reinmann, Andrew Reehorst, Mark Potapczuk, Mary Dietz, Kamel Al-Khalil, Colin Bidwell, Randy Britton, Dean Miller. Kneeling (L/R): William Wright, Thomas Ratvasky, Jaiwon Shin, Matthew Velazquez, David Anderson, Stanley Mohler. (NASA C-93-8399)

ating time if the delays were NASA's responsibility. Vincent "worked well with all of us," Reinmann emphasized, "and he worked hard at supporting and improving the IRT."[2]

In 1995, Vincent retired after "the best years of my career" and was replaced by Thomas B. Irvine. With a B.S. in civil engineering and an M.S. in engineering mechanics from Ohio State University, Irvine joined NASA-Lewis in the summer of 1982. His first assignment involved work on fractures and failures of carbon composite material that was used in jet engines. He later was assigned to the Space Station Program Office. In 1993, he joined the Aeronautical Facilities Engineering Division, where one of his first assignments was to develop a long-range plan for the IRT that included methods to increase the flow

---

[2] Reinmann to Leary, 4 July 2001; Interview with David W. Vincent by William M. Leary, 27 June 2001.

quality in the test section and settling chamber, an upgrade to the spray system with the use of modern electronic control technology, and replacement of the aging heat exchanger. When Vincent retired, Irvine applied for and secured the position of IRT facility manager.

Irvine continued Vincent's cooperative approach to the job. In part, this was a product of Irvine's personality and management style. It was also the result of a change in NASA's program management and funding stream that caused the need to create program offices separate from research organizations. In the icing world, this led to the establishment of an ad hoc icing management team to determine overall program direction. Team members included an aircraft icing program manager (first Jaiwon Shin, then Mary F. Wadel), the facility manager (Irvine), and the technology branch chief (Bond). Mark G. Potapczuk, as senior research engineer in the Icing Technology Branch, would be brought in on an as-needed basis. This ad hoc arrangement, Irvine pointed out, worked extremely well, "primarily because of the individuals involved."[3]

Irvine, like Vincent, thoroughly enjoyed his years with the IRT. Both men were impressed with the high esprit de corps among tunnel personnel. This was important because facility managers, as Irvine soon learned, had management responsibility for the facility—the "care and feeding" of the tunnel—but lacked direct responsibility for the researchers, engineers, mechanics, technicians, and other individuals who worked in the IRT and reported to different branches at NASA. But it did not matter. Everyone working in the tunnel, Irvine pointed out, had a sense that the IRT "was truly a national asset and that their work was making an important contribution to aircraft safety." When Irvine was promoted in June 1999 to the position of chief of the Facility Management Planning Office, with responsibility for all the tunnels at Lewis/Glenn, he was replaced by Susan L. Kevdzija. A graduate of Ohio Northern University with an M.S. in industrial engineering from Cleveland State University, Kevdzija joined NASA in 1990. Before taking over the IRT, she had held various engineering positions in space environmental testing, rocket and propulsion testing, and space and aeronautical facility management.[4]

Essential to the operation of the IRT since the beginning, as Zager, Vincent, Irvine, and Kevdzija emphasized, was a talented and dedicated technical support staff. The career of William P. Sexton can be seen in many ways as typical of this group, whose names rarely found their way into the reports generated by tunnel research. A Cleveland native, Sexton graduated from Brooklyn High School in 1960, then served two years in the Army. NASA employed him in June 1963 as an apprentice mechanical technician. During his four years of apprenticeship, Sexton moved into different areas at Lewis about every six months. He also attended evening classes for nine months a year that covered everything from safety to

---

[3] Interview with Thomas B. Irvine by William M. Leary, 22 September 2000.

[4] Ibid.

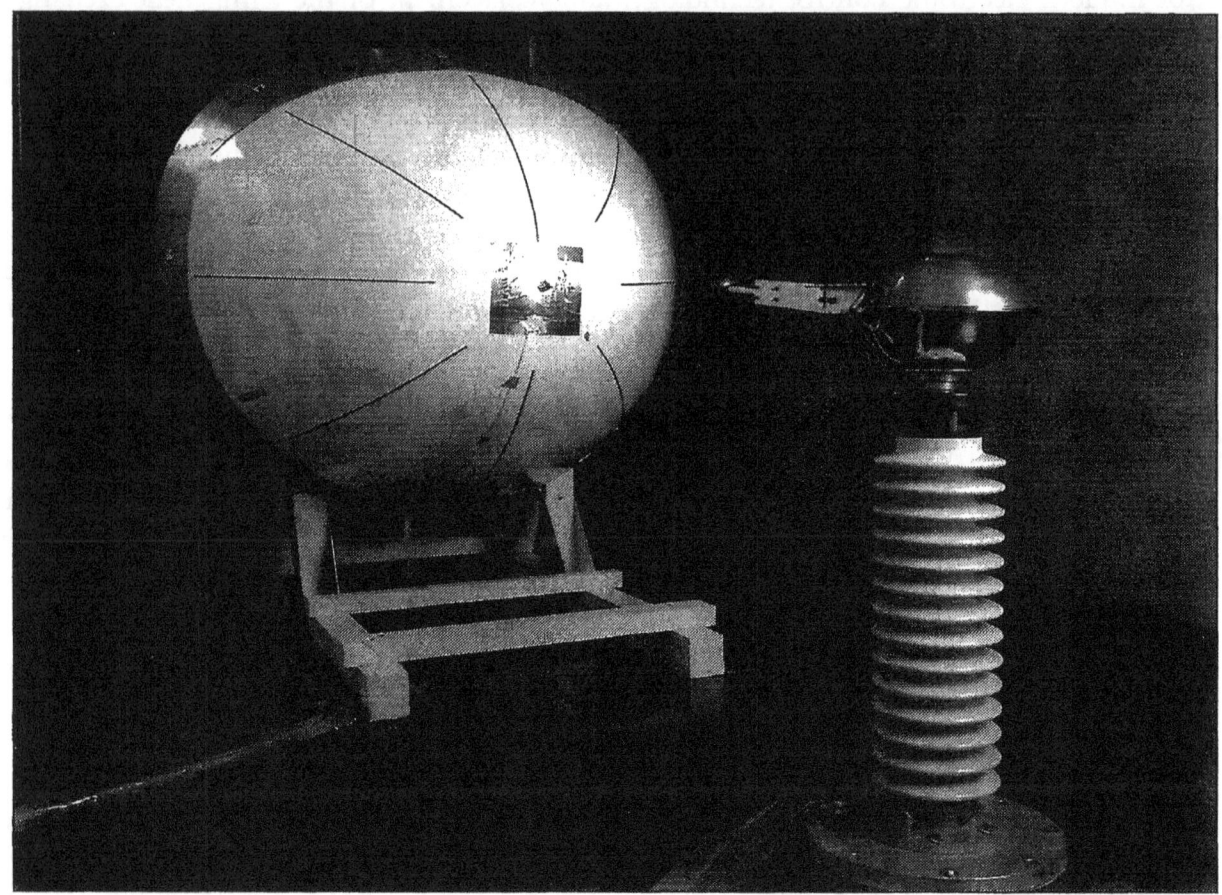

Figure 8–2. Lightning generator and radome in IRT, 1955. (NASA C-1995-4467)

physics. In 1979, he had his first tour in the IRT as a mechanical technician/tunnel operator. He spent four years on this assignment, working under Raymond G. Sotos, and was involved in all phases of tunnel operations, from maintenance to model buildup.

After a tour in the Zero-Gravity facility, Sexton returned to the IRT in 1988, one of about a dozen mechanical technicians, electronic technicians, and electricians who supported tunnel operations. These were the people who maintained the aging spray bars and controls, patched ice-damaged fan blades, integrated models into the test section, and performed thousands of other tasks that made the work of the researchers possible. Because of individuals like Sexton, who brought to their job not only a pride in workmanship but also a deep affection for the tunnel, the IRT ranked as NASA's most productive large-scale wind tunnel.[5]

---

[5] Interview with William P. Sexton by William M. Leary, 18 June 2000.

The IRT was at the center of a wide range of icing investigations during the 1990s, most of which involved cooperative efforts with industry or government agencies. In 1995, for example, NASA joined with the FAA in a program to document the ice shapes that were occurring on airfoils on aircraft currently in service and to record the effects on aerodynamic performance. The database from the Modern Airfoils Program could be used for both airfoil design and in the development of icing simulation codes.

At the beginning of the program, NASA asked aircraft manufacturers to suggest airfoils sections to be used for the tests. Researchers at Lewis selected three designs—one each for commercial transports, business jets, and general aviation. The business jet and general aviation airfoils approximated the airfoil sections found on the main wing, while the commercial airfoil section came from the horizontal tailplane. The Lewis model shop designed and fabricated three models for the IRT tests, each with a span of 6 feet and a chord of 3 feet. Technicians fabricated the commercial transport and business jet models from fiberglass, and the general aviation model primarily from laminated mahogany.

Sexton and his associates mounted the models vertically in the test section of the IRT, attaching them to an external force balance on both top and bottom, with the lower portion mounted on a turntable. The force balance was a three-component system that measured lift, drag, and pitching moment. The airfoils were then subjected to icing conditions for short (2 minutes), medium (6 minutes), and long (22.5 and 45 minutes) periods. The shorter exposures provided data on initial ice buildup and resultant aerodynamic performance, while the longer periods corresponded to FAA certification requirements for flight through known icing conditions with a failed ice-protection system (22.5 minutes) and no system (45 minutes).

Lead researcher Harold E. Addy, Jr., described a typical test run: "First, the tunnel and model were cooled to the desired temperature by operating the tunnel's fan and cooler. The fan was then brought to a stop so that a zero reading for the force balance could be taken. After restarting the fan, tunnel airspeed was brought to the desired level. At this point, force balance and wake survey measurements were made for the clean model. Then the icing portion of the test was executed for the set icing time period. Once the icing spray was terminated, force balance and wake survey measurements were taken for the iced model at the test point attitude. The model was rotated, and force balance measurements were made over a range of model attitudes. The typical attitude range was -4° to stall. Attempts to rotate the model past stall were not made due to heavy model buffeting. The model was returned to its set point attitude, and the fan was shut off. Another force balance zero measurement was made. Photographs of the ice were taken, and cuts were made in the ice so that tracings of the ice shape could be made. Normally, three tracings were made of the ice shape: one at the tunnel centerline, one 6 inches below, and one 6 inches above. Finally, the thickness of the ice at the cuts was measured using a depth gauge. The ice was then cleaned off the model, and preparations were made for the next test run."

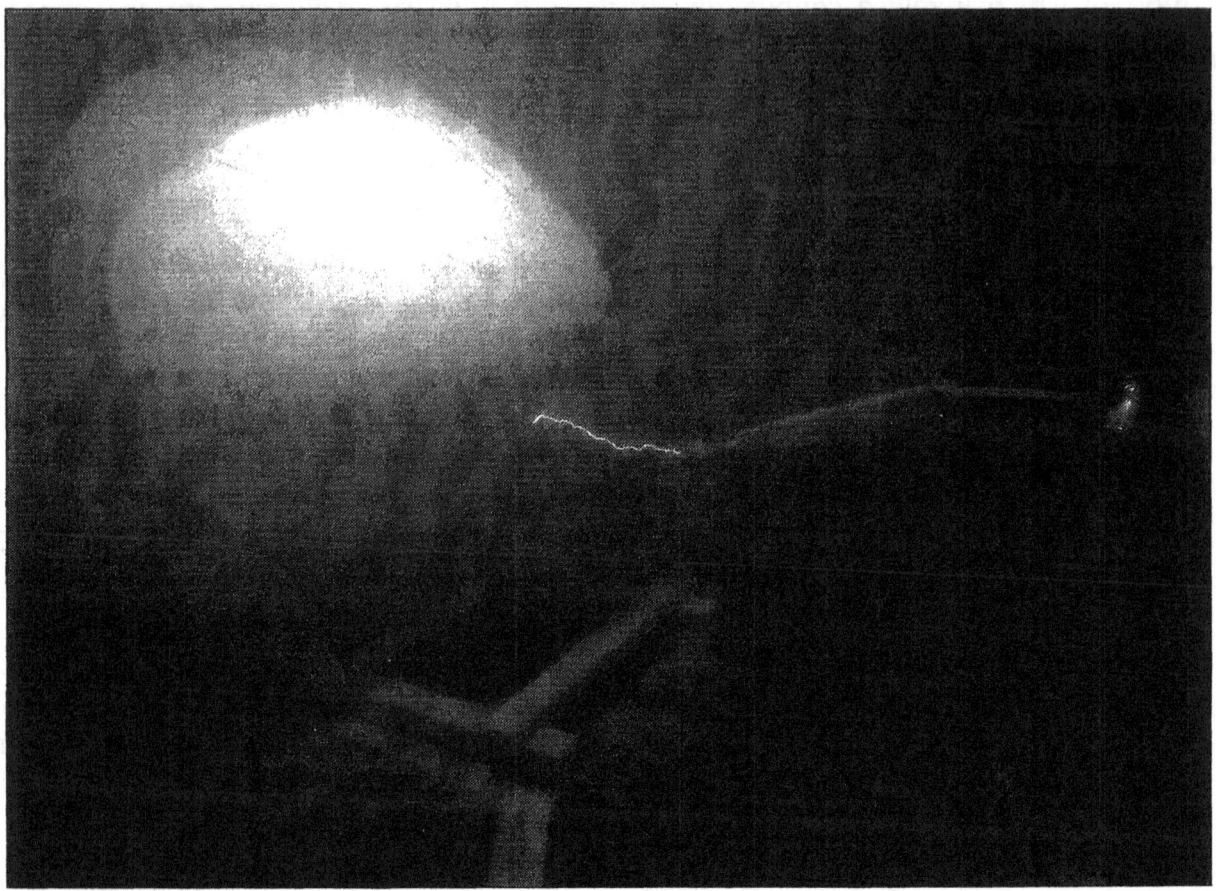

Figure 8–3. Lightning strikes on radome in IRT, 1995. (NASA C-1995-4457)

In all, there were eighty-four runs with the business jet model, forty-nine with the general aviation airfoil, and thirty-three with the commercial transport model.

Because the IRT had a relatively high level of turbulence, the Modern Airfoils Program also included tests in Langley's Low-Turbulence Pressure Tunnel (LTPT) to verify the performance measurements that had been made in the icing tunnel. The LTPT, a closed-loop tunnel with a test section that measured 3 feet wide by 7.5 feet high by 7.5 feet long, had a turbulence level of 0.1 percent or lower and could be operated at Mach numbers as high as 0.4.

For the LTPT tests, the Lewis model shop fabricated a general aviation airfoil out of solid aluminum that had a span of 3 feet and a chord of 3 feet. The model was installed horizontally in the test section of the Langley tunnel. Selected artificial ice shapes were then attached to the model between the 10- and 14-percent chords by means of flathead screws. The screws were covered with dental plaster and then smoothed to match the surrounding surface. After the instrumentation was checked, the tunnel was pressurized and the fan started. When the desired Mach number was reached, the model was rotated through a range of attitudes, using from -4° to several degrees past stall, while performance data was recorded. Researchers found that the LTPT tests supported the performance trends that had been found in the IRT tests.

The IRT data showed that exposure periods as short as 2 minutes had a significant impact on lift coefficient, especially at high angles of attack, for all three airfoils. The angle of stall also was adversely affected, particularly by glaze ice. Larger exposures to icing continued to degrade performance.

"With this data in hand," Addy concluded, "hypotheses may be formed about the physical mechanisms that play significant roles in the observed behavior of iced airfoils. Detailed investigations into these mechanisms are needed to gain physical understanding and to produce better aircraft icing tools for prediction of critical ice shapes for airfoils."[6]

The 1990s also witnessed a new effort in the decades-long search for an effective chemical compound to prevent ice from adhering on airfoil surfaces. As noted earlier, one of the earliest experiments in the NACA's first icing tunnel had tested the ice-phobic qualities of various oils, greases, waxes, and soluble compounds. The results had been disappointing. Chemist William Geer also had searched during the 1920s for an ice-phobic coating, but his tests of a number of paints and varnishes had proved equally unsuccessful.

Experiments in both the United States and abroad continued over the years. In 1969, for example, the FAA tested 100 ice-phobic coatings in the IRT. While many reduced the adhesion force of ice to the test sample, none could be used for aircraft applications. The Army's Applied Technology Laboratory also searched for a substance that might reduce the adhesion force of ice to helicopter rotor blades. In 1974, the laboratory placed an advertisement in the *Commerce Business Daily*, seeking ice-phobic coatings, and received twenty replies. Six substances were selected for tests, which were conducted by the U.S. Army Cold Regions Research and Engineering Laboratory at Hanover, New Hampshire. Two coatings showed promise—a silicone grease manufactured by GE Silicone Products Division and a silicone oil from the Dow Chemical Company. The U.S. Army Aviation Flight Activity conducted limited flight tests of the Dow coating at Spokane International Airport during the winter of 1977–78. Applied to the rotor blades of a UH-1H Huey, the Dow oil lasted for 77 minutes under test conditions. The Army concluded that the ice-phobic coating certainly had possibilities for rotor ice protection but that more testing was needed.[7]

---

[6] Addy, "Ice Accretions and Icing Effects for Modern Airfoils," NASA TP-2000-210031 (April 2000).

[7] Richard I. Adams, "Overview of Helicopter Ice Protection System Developments," Aircraft Icing: A Workshop Sponsored by the National Aeronautics and Space Administration and the Federal Aviation Administration, Lewis Research Center, 19–21 July 1978.

Figure 8–4. Super-cooled large droplet experiment with Twin Ottenberg, 1995: technicians (L/R) Timothy Hawk, David Justavick, Richard Flasig; researchers (L/R) Gene Addy, Dean Miller, and Colin Bidwell. (NASA C-1995-4010)

When NASA resumed icing research in the late 1970s, the reports on industry requirements that had been commissioned to set the research agenda for the icing branch recommended that ice-phobics be given a high priority. Reinmann was receptive. "Without doubt," he told the press in 1981, "the ideal anti-icing agent is the 'ice-phobic' one that could be permanently applied to critical surfaces, adding little weight, being low in cost and never replaced. The search for that miracle is a high-risk venture with high pay-off potential."[8]

Reinmann put together a joint program with the Army and U.S. Air Force to test ice-phobic coatings in the IRT. Using an interfacial shear rig that had been developed by operations engineer Sotos, researcher Peggy Evanich examined a number of substances, including GE silicone greases, that gave good results. Reinmann, however, became dis-

---

[8] *Lewis News*, 10 April 1981.

trustful of the shear rig measurements. "So," he recalls, "we got the bright idea of applying silicone grease to the leading edge of an airfoil and testing it in the IRT." The results, he found, "were wholly unexpected." Ice quickly built up on the airfoil, being held in place by aerodynamic forces. After the test, it was possible to move the ice "with our little finger." At this point, Reinmann concluded, "we lost faith in the concept of an ice-phobic for preventing ice buildup on an airfoil or an engine inlet."[9]

Ice-phobics, nonetheless, continued to exert its powerful lure. By the early 1990s, Reinmann had become willing to look beyond the traditional sources for ice-phobic coatings and allow amateur inventors to test their products in the IRT. This program grew out of NASA's outreach efforts, wherein NASA personnel would man booths at air shows, state fairs, and other public gatherings. It was not uncommon for the IRT people to be approached by individuals who claimed to have developed ice-phobic fluids, and Reinmann decided to give them a chance. The program would surely generate favorable publicity for NASA. And, who knows? Perhaps some did have the long-sought ice-phobic miracle liquid. He selected David N. Anderson, a talented engineer with a doctorate from the University of London, to supervise this effort.

Initially, the evaluations of the coatings were simply intended to indicate if ice did or did not stick to the treated surface. One night about every six months, groups of individuals—sometimes as many as three or four—were brought to Lewis to test their fluids in the IRT. The earliest tests often used aluminum plates or other samples supplied by the inventor and mounted in the tunnel while another experiment was in progress. Later, two aluminum poles were set up in the test section. The enthusiastic inventor was allowed to coat one of the poles with his fluid, while the other pole was left bare. The spray was then turned on to produce an icing cloud.

A colorful group of individuals usually showed up for amateur night. One time, a former racecar driver appeared in a wheelchair. After his injury in an auto crash, he had developed coatings to reduce friction in racecar engines. The coatings may have worked in racecars, but they did not keep ice from sticking. Perhaps the most memorable of the amateurs was a dentist from California, who had developed a coating for children's teeth. He arrived at the tunnel in shorts and sandals, causing raised eyebrows on the part of the staff both for the informality and inappropriateness of his dress for *icing* tests. He next produced a bottle of "magic fluid," which he proceeded to rub on his teeth. After inviting the tunnel operators to feel how slippery it was, he confidently applied the liquid to the aluminum pole. He was crestfallen to see ice build up on it.

None of the amateur-produced coatings proved viable. They would work well in a freezer, but super-cooled water droplets traveling at 100 miles per hour was a different

---

[9] Interview with Raymond G. Sotos by William M. Leary, 17 June 2000; Reinmann to Leary, 30 April 2001.

Figure 8–5. Predator fuselage in IRT, August 1996. (NASA C-96-3356)

story. The program, which lasted about three years, proved an education for the inventors and provided excellent public relations for NASA.[10]

At about the same time that these amateur-night tests were being conducted, the Air Force requested NASA to evaluate a commercially available Teflon paint as a possible ice-phobic coating. Several small chemical companies also contacted Lewis to request tests of some of their formulations. The program was further expanded to include such companies as Dow, DuPont, and 3M, who were solicited to supply candidate coatings.

For these tests, Anderson used hollow aluminum cylinders that were 3 inches in diameter. Each cylinder, mounted in the center of the tunnel test section, had three 3-inch-wide bands that were coated with the ice-phobic candidate. The coated bands were separated by 2-inch widths of bare aluminum. The cylinders were tested at temperatures of 30°, 25°, and 0°F; velocities of 100 and 250 miles per hour; water droplet median volume diameters of 15, 25, and 40 microns; and liquid water content of 0.5, 1, and 1.8 grams per cubic meter. Spray time varied from 5 to 25 minutes, depending upon how long it took to maintain a constant ice accumulation.

At the completion of the spray run, a heated aluminum plate would be used to cut a 0.25-inch gap through the ice at each coated portion of the cylinder and at an uncoated portion. The ice shapes were recorded by inserting a cardboard template into each gap and tracing the outline of the ice. A subjective assessment was made of the force required to remove the ice accumulation by striking it with a Teflon rod.

Anderson found that most of the tested coatings were disappointing. Teflon paints, for example, did no better than uncoated aluminum. Although providing a slippery surface, the Teflon also had a high pore density which likely encouraged a strong mechanical bond with the ice. A few coatings, however, especially a Dow-produced "Anti-Stick," a water-based fluorocarbon which had been developed as an anti-graffiti coating, appeared to reduce adhesion significantly.

To better evaluate the adhesion of the coatings, the next step was to try to obtain more objective data. Because Anderson lacked experience with surface coatings, he sought advice from Dr. Allen D. Reich of BFGoodrich Aerospace in the early 1990s. The promise of a low-adhesion coating remained as attractive to the boot manufacturer as it had been in the 1930s. If the adhesion of ice to the surface of Goodrich's pneumatic impulse de-icer boot could be reduced, the energy required to break the ice's surface bond also could be reduced. In addition, removal of the ice would be cleaner. Goodrich operated an icing tunnel in Uniontown, Ohio, about 40 miles south of the Lewis center, to assist their design and development program, and to provide customer support for BFGoodrich Aerospace de-icing systems. Goodrich also had an R & D center in Brecksville, Ohio,

---

[10] Sexton interview; interview with David N. Anderson by William M. Leary, 23 September 2000.

Figure 8–6. Glycol pump used for Predator "weeping wing" tests in the IRT, August 1996. (NASA C-9-3380)

which was 15 miles east of NASA Lewis. The proximity of Goodrich facilities and their extensive corporate experience in surface-coating research made the company a natural source of advice. Reich was Goodrich's in-house expert on surface coatings at that time and had put together a bench-top shear rig and test procedure for measuring the adhesion of ice to various surfaces. The discussions between Anderson and Reich resulted in an informal cooperative working relationship in which Reich would test some of the coatings Anderson had previously tested in the IRT in the Goodrich shear rig in Brecksville.[11]

Reich tested the most promising coatings from the IRT experiments. Freezing static ice in a thin layer (0.07 inches) over the candidate coating that had been applied to a stationary platform did this. A gear motor then drove a movable pedestal. The force

[11] Anderson to Leary, 11 June 2001.

required to remove the ice was measured with a load cell. Reich could calculate adhesion force with a simple mathematical formula. In these tests, the Dow coating also came out on top, displaying fairly repeatable adhesive strength that increased only slightly with a number of ice removals.

While the Goodrich tests were encouraging, Anderson knew that impact ice had different physical characteristics than static ice. In an attempt to obtain quantifiable results with impact ice in the IRT, he employed in the next series of tests a shear rig that had been developed at the University of Akron by R. L. Scavuzzo and M. L. Chu. Based on the earlier NASA device, it used a 1-inch-diameter cylinder that would be coated with test material and mounted inside a thin sleeve. A window that was cut from the sleeve exposed a portion of the cylinder to the icing cloud. Ice accreting on the sleeve bonded the cylinder to the sleeve at the window. Once the ice had accreted, the cylinder-sleeve assembly was removed and placed on a stand. The cylinder was then forced through the sleeve by a hydraulic press acting through a load cell. The force required to free the cylinder was recorded on a strip chart.

Anderson soon found that the mechanism suffered from two major problems. Ice sometimes caused adjoining cylinder-sleeve assemblies to stick together, making disassembly difficult. Also, the cylinder often separated from the sleeve when removed from the test stand, especially with low adhesion coatings. As a result, no measurements could be obtained. In the end, Anderson considered the data that he obtained from the shear rig tests to be unreliable.

NASA technicians managed to devise a test stand that permitted shear-force measurements to be made in the tunnel without disturbing the coating specimens. Two coatings stood out in the subsequent tests. Both the Dow "Anti-Stick" coating and a fluoropolymer-based DuPont coating reduced adhesive strength by 30 to 50 percent. These results were confirmed in the Goodrich tunnel. At this point, NASA concluded its work on the ice-phobic project, leaving Goodrich and the chemical companies to develop further a practical low-adhesive coating.[12]

One of the most innovative uses of the IRT during the 1990s came with tests of the effects of lightning strikes on radomes. Because radomes were constructed from non-electrically conducting materials, they were vulnerable to being punctured or shattered by lightning strikes. To minimize this danger, radomes initially had been fitted with an arrangement of solid metal bars, called diverters, that would conduct the lightning safely to the airframe. Newer protection devices—segmented diverters—accomplished the same purpose but caused less interference with radar performance. Verification of the

---

[12] Anderson interview; David N. Anderson and Allen D. Reich, "Tests of the Performance of Coatings for Low Ice Adhesion," NASA TM 107399 (1997).

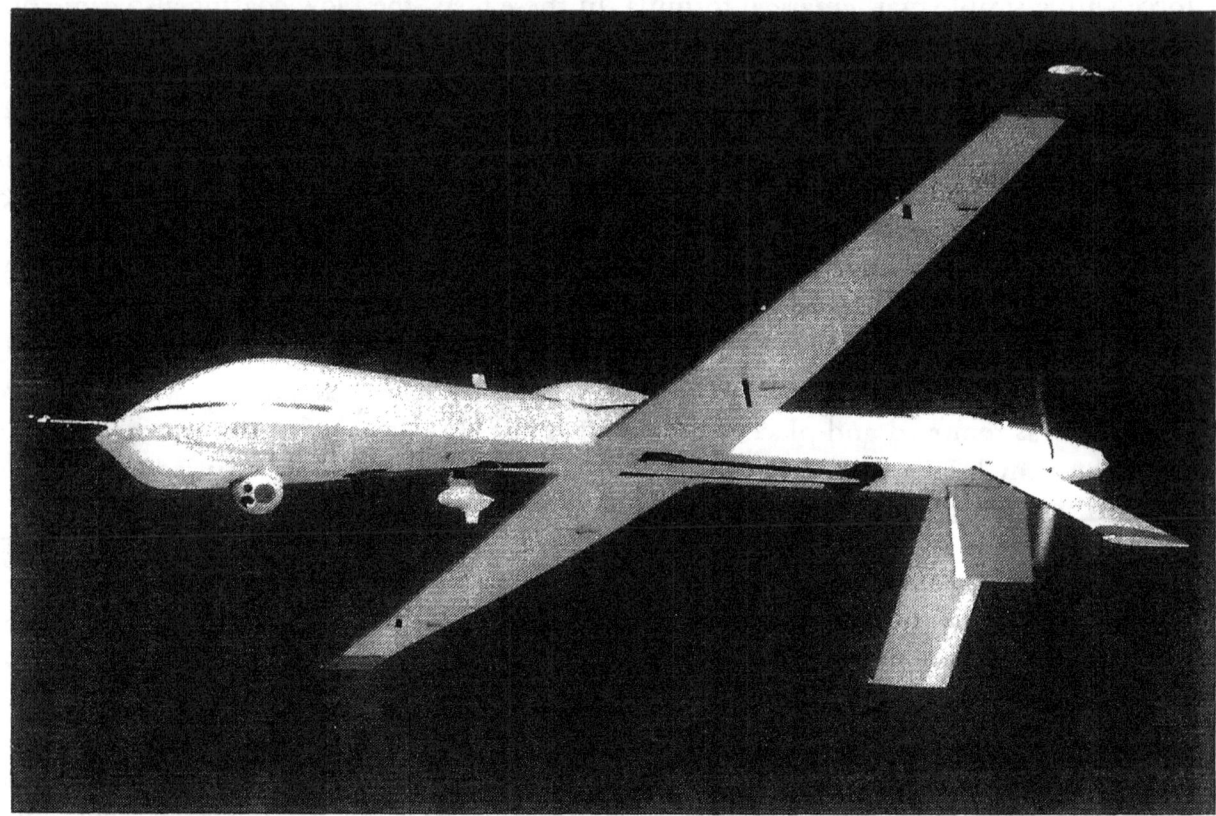

Figure 8–7. Predator icing test, Duluth, MN. (NASA C-97-01251)

effectiveness of these devices was accomplished by simulated lightning strikes, usually conducted within indoor high-voltage laboratories. Even after having passed these certification tests, however, the radomes continued to be vulnerable to lightning strikes, suggesting that atmospheric conditions—rain, ice, airflow—might influence the performance of the designs.[13]

Late in 1993, a consortium of government agencies and commercial companies in the United States and Europe formed a joint program on the improvement of lightning and static protection of radomes. One of the objectives of the multi-year program was to test the segmented diverter system under simulated environmental conditions. The FAA sought the assistance of NASA-Lewis to conduct these tests. After an inspection of the IRT in February 1995 determined that the tunnel could accommodate the experiment, the tests were scheduled for October 1995.

---

[13] Irvine to Michael S. Glynn, program manager, FAA Technical Center, 20 September 1995, records of the icing branch, Glenn Research Center.

The lightning generator of the tests would be provided and operated by Lightning Technologies, Inc. The 500-kilovolt Marx-type impulse generator, standing 8 feet tall on a 3-foot metal base, produced high-voltage impulses reaching a peak amplitude of 500 to 700 kilovolts in times selectable from 1 to 50 microseconds. This voltage could flash over between 0.5 and 1.5 meters of air. For the IRT tests, it was installed in the diffuser section of the tunnel, downstream from the usual test section. This larger area, in which airstream velocities of up to 200 miles per hour could be reached, was necessary to contain the lightning generator and radome models.

IRT technicians went to great lengths to ensure that the experiments did not damage the tunnel or its sensitive equipment. Just to be sure that no unforeseen effects occurred, the generator was operated at low voltage levels when tests began, and then it was slowly increased to normal test voltage amplitude. As it turned out, the visually impressive weeklong tests took place without incident. A good deal of useful data was recorded to assist the FAA in the development of new certification standards.

Urgent projects generated by the military services always commanded a high priority in the IRT. One such program during the 1990s involved the development of an ice-protection system for the Predator Unmanned Aerial Vehicle (UAV). Designated the RQ-1 by the Air Force, the Predator was manufactured by General Atomics Aeronautical Systems, Inc., without ice protection. It appeared in the mid-1990s as an effective reconnaissance platform that could be operated over high-risk areas without the danger of human loss. Equipped with four state-of-the-art sensors, Predators could fly over targeted areas for nearly 40 hours. Essentially a powered glider, the air vehicle was a mid-wing monoplane with a slender fuselage that housed the payload and fuel. It featured a high aspect ratio wing and an inverted-V tail.

A prototype Predator participated with great success in the annual air defense exercises that were held in the southwestern United States in the spring of 1995. Its first operational deployment came in July 1995. Based in Albania, the Predator flew surveillance missions over the Balkans until November 1995. A second European deployment took place in March 1996, with the Predators being based in Hungary.

These early missions proved successful, although the military became concerned over the vehicle's lack of ice protection. The Predator's natural laminar flow wing was sensitive to any form of contamination. Rain, for example, would cause it to descend. It could not operate at all in an icing environment, which limited its usefulness.

The Department of Defense's Joint Project Office called upon NASA for assistance in the spring of 1996. The Air Force quickly fabricated three models for testing in the IRT. Researcher Andrew L. Reehorst began the test program on 5 August. He first placed in the tunnel a truncated half-fuselage model of the Predator that was complete with inlets and antenna. The main purpose of these experiments was to identify a suitable location for a Rosemount ultrasonic ice detector. The tests showed that the planned loca-

Figure 8–8. Ice on communications antenna of Predator, Duluth, MN. (NASA C-97-01254)

tion for the ice detector was masked by vortices and would not have worked. A different location was then selected and tested.

The next phase involved tests of two ice-protection systems on two wing models. Between 6 and 9 August, Reehorst ran experiments with a "weeping wing" system that was being produced by Aerospace Systems & Technology (AS&T) of Salinas, Kansas. First developed in Great Britain during World War II and known as the TKS system, the "weeping wing" came about because the Royal Air Force's Bomber Command had found that wings with pneumatic boots were catching on the cables of anti-aircraft balloons and suffering extensive damage. The answer was an armored wing to cut through the cables. But this type of wing required ice protection. Three British companies rose to the challenge. Tecalemit (T) provided pumps, Kilfrost (K) supplied de-icing fluid, and Sheepbridge Stokes (S) manufactured alloy strips that were inserted into the wings and distributed the fluid.

The TKS system, with modifications, was used extensively in Great Britain in the postwar years. In the early 1980s, the University of Kansas, under a grant from NASA-

Lewis, tested the system in the IRT. Using two general aviation airfoils, the Kansas researchers examined two distribution methods—one that employed a stainless steel mesh and the other made from porous composite material.[14]

AS&T was formed in 1990 to build leading-edge TKS structures for general aviation aircraft. It grew rapidly and soon acquired the British manufacturers of the product. The TKS system tested on the Predator featured porous, laser-drilled titanium panels to distribute a glycol-based fluid to wings and horizontal and vertical stabilizers. A small electrically driven pump and storage tank completed the system, which weighed only 40 pounds (without liquid) and consumed little power.[15]

Between 12 and 16 August 1996, Reehorst used the second wing section for experiments with the more innovative electro-mechanical de-icing system that was under development by Innovative Dynamics, Inc., (IDI) of Ithaca, New York. IDI's Sonic Pulse Electro-Expulsive De-icer (SPEED) had evolved from earlier work on Electro-Impulse De-Icing (EIDI) (see chapter five). As with EIDI, SPEED featured actuator coils that were placed behind the leading edge of the airfoil and applied impulsive loads directly to the aircraft skin. The rapid electrical acceleration debonded the ice, which was carried away in the airstream. With major improvements in the activator coil and electronics, SPEED had performed well in earlier IRT tests, shedding ice as thin as 0.050 inches. As with the TKS system, SPEED offered a low-weight, low-power system of ice protection.[16]

Reehorst obtained excellent results with both the TKS and SPEED systems. Although SPEED was the more technologically advanced method of ice protection, the Air Force opted for the more proven TKS system. Flight tests of the Predator with the "weeping wing" took place at the Army's test facility at Duluth, Minnesota, during the winter of 1996–97. This was followed by a second series of flight tests in the winter of 1997–98, leading to a decision to incorporate the TKS system on the "Block 2" (RQ-1B) series of Predators.[17]

The need for the ice-protected Predators was made abundantly clear to the Air Force when a RQ-1A crashed near Tuzla Air Base, Bosnia, due to icing on 18 April 1999. The Department of Defense removed the aircraft from the Balkans in the fall of 1999, due to the dangers from

---

[14] On the University of Kansas tests, see Reinmann, R. J. Shaw, and W. A. Olsen, Jr., "Aircraft Icing Research at NASA," NASA TM 82919 (1982).

[15] David K. Henry, "TKS Ice Protection," 4 May 2001. The author is indebted to Mr. Henry of AS&T for putting together this historical account of the TKS system.

[16] Interview with Andrew L. Reehorst by William M. Leary, 22 September 2000; Reehorst to Leary, 4 May 2001; Innovative Dynamics, Inc., "Sonic Pulse Electro-Impulsive De-Icer." The author is indebted to Mr. Joseph J. Gerardi of IDI for information on the SPEED system.

[17] Reehorst to Leary, 4 May 2001.

Figure 8–9. (L/R) Judith VanZante, Richard Ranaudo, and Thomas Ratvasky in front of a Twin Otter during the tailplane icing program, 1997. (NASA C-97-3962)

icing. In April 2001, the first ice-protected RQ-1Bs commenced operations from Macedonia's Petrovec airport, flying surveillance missions over the southern Balkans.[18]

NASA's Predator ice-protection program, Reehorst observed, was unique because of the speed with which the system went from concept to reality. The project again demonstrated the icing branch's ability to respond promptly to urgent military requirements.[19]

As the IRT approached its 50th anniversary, it was clear that the heavily used tunnel needed major modifications if it were to continue as a world-class research facility in the 21st century. In 1993, NASA-Lewis contracted with Sverdrup Technologies for a study of what changes would be necessary to upgrade the IRT. David Spera, who earlier had contributed the idea of changing the pitch of the IRT's fan blades to increase tunnel

---

[18] American Forces Press Service, "RQ-1 Predator Accident Report Released," 23 December 1999, and "Predator Demonstrates Worth over Kosovo," 21 September 1999.

[19] Reehorst interview.

speed, was the principal investigator for Sverdrup. He suggested that significant improvements be made in flow quality, water spray uniformity, load balance accuracy, and the drive fan system. These modifications would allow the effects of icing and de-icing on lift and drag to be measured more accurately under a wider range of conditions. He estimated that the project would cost $20 million.[20]

Spera sent his plan to facility manager Vincent, who submitted it up the NASA chain-of-command as a national facilities study project. As it turned out, the Spera plan became the basis for an even more extensive tunnel modification program that had been developed by Irvine while he was working in the Aeronautical Facilities Engineering Division. Work began late in 1993 with structural repairs of the IRT's pressure shell and supports.[21]

Work on the spray bar system got underway in September 1996. Under the existing arrangement, the uniform icing cloud covered approximately 30 percent of the test section area. The goal of the new system was to cover at least 60 percent. Relying upon empirical data from the existing system, the design team concluded that adding two spray bars would produce the desired icing cloud. With a total of ten bars, the new system could produce spray from up to 540 locations. In addition, a second water line that was connected to each nozzle holder permitted rapid changes between two sets of liquid water content/median volume diameter test conditions.

Two other important spray system changes were based on the experiences of the Boeing Research Aerodynamic Icing Tunnel (BRAIT). Each spray nozzle in the Boeing facility had a solenoid valve for electrical control. With this valve, it was possible to achieve exact air and water set points, and stabilized flow within 15 seconds. The settling time for NASA's existing system, which used a supply valve located upstream on the water header, was between 30 and 90 seconds, with a typical pressure overshoot of 10 to 15 percent of the desired set points. NASA concluded an agreement with Boeing for the use of their superior controls for the new IRT spray bar system.[22]

Boeing's experience also influenced the selection of an aerodynamic shape for the spray bars. The flow quality in the Boeing tunnel was superior to that in the existing IRT in part because of Boeing's use of aerodynamically shaped spray bars. NASA's design team adopted a modified NACA 0024 airfoil shape for the new spray bars in hope of improving flow quality.

---

[20] David W. Vincent, "Upgrade of the Icing Research Tunnel (IRT) at the Lewis Research Center," 9 July 1993. The author is indebted to Mr. Spera for a copy of this document.

[21] *Lewis News*, 3 December 1993.

[22] Thomas B. Irvine, John R. Oldenburg, and David W. Sheldon, "New Icing Cloud Simulation System at the NASA Glenn Research Center Icing Research Tunnel," NASA TM-1999-208891 (June 1999); George E. Cain to Leary, 3 November 2000.

Figure 8–10. Calibration tests in the IRT, 1997. (NASA C-1997-1181)

Construction of the new spray bar system took until January 1997. At this point, Robert Ide and Jose Gonzalez assumed responsibility for icing cloud and aerodynamic calibration. Gonzalez showed that test section aerodynamic flow quality was not measurably different with the new system, with the exception of an overall decrease in turbulence. Ide's cloud calibration produced a set of calibration curves that would be used by tunnel operators to set spray bar conditions to achieve the desired icing cloud characteristics. He also was able to achieve a uniform icing cloud covering 60 percent of the test section.

The IRT reopened with the new spray bar system in May 1997. All went well at first, but then the facility became "unstable," that is, icing cloud uniformity became impossible to achieve. Facility manager Irvine, who had replaced Vincent in March 1995, promptly launched an investigation, checking the aging heat exchanger, new spray bars, and everything else that he could think about. "Nothing," Irvine recalled, "could be found." Finally, he brought in Edward J. Zampino from the Safety Office. Zampino developed a fault tree analysis and systematically investigated all possible root causes of the problem. He never managed to identify the reason for the instability; although, a good deal was learned about the operation of the tunnel during the investigation. After four or five frustrating months, the problem simply cleared up. It then took another four months to recalibrate the tunnel. When this tedious job was finished, the IRT's staff faced months of overwork to complete critical tunnel projects that had been delayed. Irvine recalls this as a difficult period in the operation of the facility when it was hard to keep up morale.[23]

With the IRT scheduled to shut down in the spring of 1999 for replacement of the heat exchanger, work in the heavily used facility intensified during the first part of the year. January began with a continuation of the Tailplane Icing Program. "We were looking to gain a broader understanding of the aero-performance effects of icing on a business jet horizontal tail," researcher Ratvasky pointed out, "and to develop test methods using sub-scale models with artificial ice shapes to obtain useful data for full-scale comparison." With the support of engineers from Learjet, Ratvasky ran 45.1 hours of icing tests of a full-scale horizontal tail of a Lear business jet that produced realistic ice shapes. Molds were made of two shapes, which were subsequently made into casts that were used for full-scale empennage tests in the 40 x 80-foot wind tunnel at the Ames laboratory.[24]

David Loe of Bell Helicopter Textron followed Ratvasky in the tunnel. Loe spent 79 hours in the IRT, collecting ice shapes on the empennage of the Bell Agusta 609 tilt rotor.

---

[23] Irvine interview.

[24] Ratvasky to Leary, 24 July 2001; Icing Research Tunnel, "1999 Schedule and Model Information Report," copy in the files of the icing branch, Glenn Research Center.

Figure 8–11. Radome of the MC-130E Combat Talon in the IRT, 1999. (NASA C-99-233)

The casts resulting from these shapes would be attached to the innovative civil tilt rotor for the Zero-G pushover tests that the FAA now required as part of icing certification.[25]

The month ended with 27.9 hours of tests that had been requested by the U.S. Air Force, which had been experiencing difficulties with the hot air anti-ice system used for radomes on MC-130 Combat Talons. The radome for the aircraft used by Air Force special forces had been redesigned from the one on standard C-130 transports in order to accept new terrain following radar. When the anti-ice system had been turned on, however, the radome had a tendency to overheat and implode. In three inflight tests behind an ice tanker, a modified anti-ice system had failed to meet specifications. Rather than remove the aircraft from operational status for further tests, the Air Force decided to continue testing in the IRT.[26]

With the assistance of personnel from Lockheed-Martin, IRT technicians built a rig to hold a Combat Talon radome for icing experiments. It turned out that the anti-ice valving system was putting out too much hot air, causing cracks in the radome. The solution to the problem was simply to adjust the valving system so that enough hot air was generated to keep ice off the radome without cooking it.[27]

February brought a month-long series of droplet impingement studies that were conducted in partnership with Wichita State University (WSU). A total of 147.3 hours of tunnel time was devoted to experiments with models of several wing sections and an inlet that had been selected following consultation with the aircraft industry. The purpose of the tests was to expand the droplet impingement database that was used to design ice-protection systems on new aircraft and to validate ice-prediction computer codes.

The technique used for the droplet impingement experiments was a modern version of the dye tests of the 1950s. Areas of interest on a model would be covered with blotter strips, then sprayed with a dye of known concentration. Measuring the amount of dye on each of the strips would determine the rate of water impingement for the region. The more recent version, however, used a high-performance spray system that had been developed at Wichita State. Mounted on the IRT spray bars, it generated short, stable sprays of less than 10 seconds. Unlike the previous red dye, the WSU system sprayed blue dye. This was because the measurement of dye concentration on the blotter strips was accomplished by bouncing a laser beam off the surface of the strips. The intensity of the reflected light could be measured and correlated to the amount of dye on the strips. The blue gave the best reflective results for the red laser beam.

Although the 1950s system, as noted earlier, was both labor intensive and time consuming, the newer version also posed challenges for the experimenters. "The impingement

[25] Leary telephone interview with Roger E. Aubert, 24 July 2001.

[26] "Anti-Ice Test Plan," 25 November 1998, icing branch files.

[27] Sexton interview.

Figure 8–12. Icing tests of Sikorsky S-92 inlet in the IRT, 1998. (NASA C-98-1348)

tests are rather lengthy and grueling," researcher Colin S. Bidwell explained, "because of the time required to install the customized spray system and the number of models tested. The spray system takes several weeks to install and calibrate. A typical model is tested for one to three days and involved [in] aero-testing (for model surface pressures) and impingement tests for various attitudes and spray droplet diameters. Each of the conditions requires three to five repeats to generate a useful statistical average. A typical model might generate 300 blotter strips."

The 1999 impingement experiments were the third set of tests that had been conducted since 1985. Funded by the FAA and Boeing, and with manpower and model support from a variety of manufacturers, the program was scheduled to conclude with the final series of tests in September 2001.[28]

---

[28] Bidwell to Leary, 23 July 2001; Michael Papadakis, Marlin D. Breer, Neil C. Craig, and Bidwell, "Experimental Water Droplet Impingement Data on Modern Aircraft Surfaces," AIAA Paper 91-0445 (January 1991).

March saw 44 hours of tests performed on the wing section of the Bell 609 tilt rotor—this time to demonstrate the operation of the de-icing system and to collect ice shapes for tests of performance degradation between de-icing cycles. In addition, experiments were conducted with the horizontal tail sections of two business jets. Over 14 hours of tunnel time involved icing tests for the certification of the Sino Swearingen SJ30-2, a seven-place, twin jet engine business aircraft. Icing certification tests of the Raytheon Premier 1, another twin jet business airplane, required 72 hours of tunnel time. The low-energy ice-protection system for the Premier 1, developed by Cox & Company of New York City with support from Glenn's Small Business Innovative Research funding program, featured an imaginative combination of thermal anti-icing for the leading edge of the airfoil and electro-mechanical expulsion de-icing for the area past the heated edge. Further along in the certification process, the Premier 1 went on to receive icing approval from the FAA in May 2001. All tests in March involved reimbursable funding—with $130,354 for Bell Textron, $54,556 for Swearingen, and $215,242 for Raytheon.[29]

The period between March 31 and April 20 witnessed 128.3 hours of testing that was designed to provide rotorcraft airfoil ice-accretion data for correlation with the LEWICE code and to evaluate de-icing systems for application to rotorcraft airfoils. The National Rotorcraft Technology Center (NRTC) of the Rotorcraft Industry Association had initiated in 1997 a project to develop rotor blade ice-protection systems for conventional helicopters and tilt-rotor aircraft. NRTC researchers began by evaluating an extensive list of ice-protection methods. Ice-phobic material, they concluded, lacked the ability to shed ice and to withstand rain and other sources of erosion. They then wrote generic specifications for selecting manufacturers to have their ice-protection systems tested. Three suppliers responded with an electro-expulsive and two electro-thermal systems. In April 1999, the systems were installed on a Sikorsky SC2110 airfoil (15-inch chord) in the IRT. Under the direction of Robert J. Flemming, chief of experimental aeromechanics and icing at the Sikorsky Aircraft Corporation, the NRTC team acquired ice-accretion and de-icing data for the test models, along with airfoil lift, drag, and pitching moment measurements. The tests revealed that the ice-shedding capabilities of the electro-thermal systems were superior to the electro-expulsive system for rotor blade applications. As a result of the IRT tests, Sikorsky selected an electro-thermal system for its S-97 helicopter.[30]

---

[29] IRT, "1999 Schedule."

[30] Flemming to Leary, 23 July 2001, enclosing "Rotor Ice Protection Systems Development Tests in the Icing Research Tunnel."

Figure 8–13. Bell Augusta 609 tilt rotor empennage during icing tests, 1999: (L/R) technicians Tom Lawrence and Michael Lupton. (NASA C-99-204)

In all, the IRT operated for an impressive total of 639.2 hours and generated $726,008 in reimbursable funds before the tunnel was released to East/West Construction on 10 May 1999.[31]

While the Carrier heat exchanger had been an engineering marvel when installed in the early 1940s, it had deteriorated badly over the years. Leaks were commonplace. The system held 60,000 pounds of R-12 refrigerant, plus a small reserve. It would lose the entire supply of R-12 over a year's time, and the cost of the refrigerant had risen from 50 cents to $3 a pound. Technicians controlled the leaks by pinching off tubes, which meant that the temperature of the air flowing over the exchanger became less uniform. Maintaining the Carrier unit was accounting for approximately 100 hours of tunnel downtime per year by the 1990s. "Although a new heat exchanger made sense," facility

---

[31] IRT, "1999 Schedule."

manager Irvine recalled, "there was still the thought that it would mean tearing apart a national treasure that was operational." Nonetheless, it had to be done.[32]

Once the decision was made, everyone connected with the project worked hard to ensure that they got it right. Again, as Irvine pointed out, "the Boeing experience was indispensable." The BRAIT heat exchanger was a state-of-the-art flat face unit that measured 19'8" by 34'8". Using environmentally friendly R-134a refrigerant, it was located between corners "C" and "D" in the tunnel. (Boeing used the designation #3 and #4.) Air from the fan flowed around corner C, passed over the heat exchanger, then exited via corner D toward the spray bars and test chamber.

Dr. Seetharam H. Chintamani, a talented Boeing aerodynamicist, recognized that the design of the turning vanes in corner C was crucial. In order for the heat exchanger to maintain temperature uniformity of +/- 1°F in the test section, the manufacturer needed an approach velocity of no more than 15 feet per second, with a maximum variation of +/- 10 percent. Using a combination of computational fluid mechanics and empirical methods, Chintamani selected various turning vane designs for one-fifth-scale model tests. He found that a simple cambered flat plate with a 90° turning vane offered the desired flow qualities; however, it suffered from periodic flow separation. Chintamani came up with a unique leading-edge addition that solved the problem. He also fine-tuned the turning vanes at the D corner to ensure velocity uniformity for the air as it approached the spray bars.[33]

The Boeing innovations produced a far superior distribution of temperature and flow quality than was found in the IRT. "Without the Boeing model and experience," Irvine pointed out, "it is unlikely that NASA would have had the engineering insight—or courage—to go ahead with the redesign of the heat exchanger."[34]

NASA also selected the flat-face heater exchanger, configured as two side-by-side units, each with its own coolant supply of R-134a refrigerant and return pipes. Because the existing heat exchanger and surrounding steel structure were so closely connected, this meant that the Carrier unit could not be replaced without major structural modifications. The width of the tunnel duct between corners C and D had to be increased from 29.2 to 49.2 feet. This added width necessitated new turning vanes in both corners. There would be sixteen aerodynamically designed vanes at each corner. The shape of the vanes in corner C slowed down

---

[32] Sexton interview; Irvine interview.

[33] Chintamani to Leary, 29 September 2000; Chintamani and R. Steven Sawyer, "Experimental Design of the Expanding Third Corner for the Boeing Research Aerodynamic Icing Tunnel," AIAA paper 92-0031, presented at the 30th Aerospace Sciences Meeting and Exhibit, 6–9 January, Reno, NV. The author is indebted to Dr. Chintamani for a copy of this paper and other material relating to the design of the Boeing tunnel.

[34] Irvine interview.

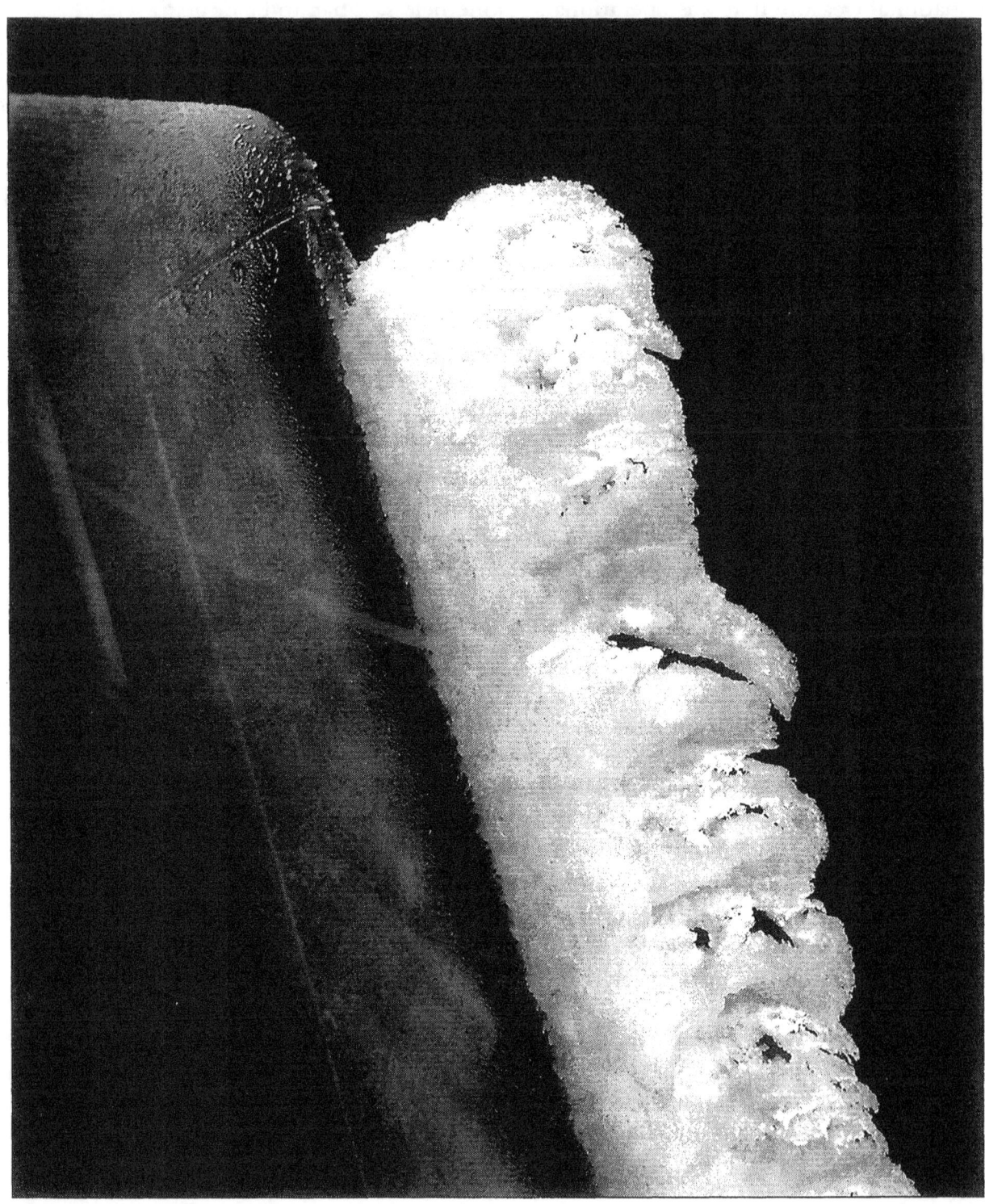

Figure 8–14. Ice shapes, 1999. (NASA C-99-191)

the air flow as it passed over the heat exchanger, while the vanes in corner D speeded up the flow as it approached the spray bars. Unlike the Boeing tunnel, which used the Chintamani-designed closely spaced curved steel plates as turning vanes, the IRT's designers selected thick cambered airfoils, both for reasons of accessibility and aerodynamic performance.[35]

A design contract for the modification was awarded to Aero System Engineering, Inc., of St. Paul, Minnesota. The company developed project design criteria, produced final design drawings and specifications for the new C–D leg of the tunnel, and developed subcontractor specifications. Cloudy & Britton, Inc., of Mountlake Terrace, Washington, the builder of the BRAIT heat exchanger, contributed the preliminary design for the IRT replacement unit. Frigid Coil/Imeco, Inc., of Santa Fe Springs, California, did final design and fabrication of the two side-by-side heat exchangers.[36]

Irvine vividly recalls the morning of 10 May 1999, when demolition work began. A large hydraulic cutter ("the jaws of death") was brought in to cut up the old heat exchanger. It was an emotional experience to witness the destruction of a national treasure. He remembers thinking, "I hope that we were doing the right thing."[37]

The construction went smoothly, and the tunnel was ready for reactivation tests on 23 November 1999. Cloud calibration proved a major challenge for Robert Ide. Because of the changes made to the quality of the airflow entering the spray bar section, hundreds of nozzles had to be repositioned in order to obtain maximum uniformity in ice accretion in the test section. As it turned out, the improved flow quality of the air approaching the spray bars had a small negative effect on the uniformity of the icing cloud, as the turbulence produced by the previous system had actually helped to mix the water droplets into a more uniform cloud. Tunnel activation, with its startup challenges, took 571.1 hours. Finally, all was ready and the tunnel went back on line in July 2000. As of 17 November 2000, the modified facility had operated for 978 hours without incident.[38]

Icing branch chief Bond has pronounced himself "very satisfied" with the construction results. In 1994, a peer review had pointed out that poor airflow quality was one of the tunnel's major drawbacks. This had now been remedied. The airflow qualities, Bond noted, had been significantly improved, as had temperature uniformity.[39]

---

[35] Irvine, Susan L. Kevdzija, David W. Sheldon, and David A. Spera (Dynacs Engineering Company), "Overview of the Icing and Flow Quality Improvements Program for the NASA Glenn Icing Research Tunnel," NASA TM-2001-210686 (February 2001).

[36] Ibid.

[37] Irvine interview.

[38] Irvine et al., TM-2001-210686.

[39] Douglas J. Silva to Reinmann, enclosed copy of the NASA-LeRC Icing Research Peer Report, 23 March 1994, History Office, GRC; Bond interview.

Figure 8–15. Demolition of Corner D and original Carrier heat exchanger, 1999.
(NASA C-99-1452)

Bond sees a bright future for the IRT. Outside requests for use of the facility are greater than ever, and there is a backlog of nearly two years. A decision by Boeing to shut down its icing tunnel will only increase the competition for IRT time. Pressure on the aircraft industry to develop new regional transports will generate significant work for the tunnel. Demands to investigate large droplet icing continue. The rotorcraft industry's search for ice-protection systems means work for the IRT, as does all-weather requirements for unmanned reconnaissance vehicles. NASA will continue to develop new protection technologies and improve computer prediction codes. Without a doubt, the IRT will remain the nation's premier icing research facility for the foreseeable future.

Essay on Sources

A central body of records on the icing research that has been conducted at the NACA/NASA Lewis-Glenn laboratory does not exist. While seeking to identify material to support this study, historians and archivists at the Glenn Research Laboratory searched through stored record collections, as well as long-unopened file cabinets at the Icing Research Tunnel, in an effort to locate pertinent documents. They uncovered only a handful of letters, memoranda, and other primary sources, which are now located in the History Office at Glenn. Inspection of the holdings of the History Office at NASA Headquarters also revealed only a limited collection of documents bearing on icing research. Icing material in Record Group 255 at the National Archives and Records Administration proved neither easy to identify nor especially helpful. No doubt, additional material is buried in the collection. The best documentary source on icing research is in file RA 247 at the Langley Historical Archive, Floyd L. Thompson Technical Library, Langley Research Center, Hampton, Virginia. This rich collection, however, covers only icing research prior to 1944.

Secondary sources on icing research are few and far between. There is a fine essay by Glenn E. Bugos, "Lew Rodert, Epistemological Liaison, and Thermal De-Icing at Ames," in Pamela E. Mack (ed.), *From Engineering Science to Big Science: The NACA and NASA Collier Trophy Research Project Winners* (NASA SP-4219, 1998), pp. 29–58. Bugos, however, has little to say about events in Cleveland due to the focus of his study. There is also a chapter on icing research in George W. Gray, *Frontiers of Flight: The Story of NACA Research* (New York: Alfred A. Knopf, 1948), but, again, the spotlight is on Rodert. Michael H. Gorn kindly supplied me with a draft chapter on icing flight research from his now published study, *Expanding the Envelope: Flight Research at the NACA and NASA, 1915–1988* (Lexington: University Press of Kentucky, 2001), which proved most helpful. Finally, Virginia P. Dawson's excellent *Engines and Innovation: Lewis Laboratory and American Propulsion Technology* (NASA SP-4306, 1991), not only provided the context in which icing research took place at Lewis, but also briefly discussed the construction of the IRT and its subsequent use.

Two published sources in icing that are aimed at the manufacturers, operators, and regulators of aircraft contain much useful information, much of it technical in character. Known as "the icing bible," Federal Aviation Administration Technical Center, *Aircraft Icing Handbook* (DOT/FAA/CT-88/8-1, 3 vols., 1991), is a rich source of material on the topic. Also helpful is Flight Safety Foundation, *Protection Against Icing: A Comprehensive Overview*, a special issue of *Flight Safety Digest*, vol. 16 (June–September 1997).

With only sparse documentation in terms of letters, memoranda, journals, and the like, the story of icing research at the Cleveland laboratory has had to rely primarily on technical reports, plus oral history and correspondence with people who were there. Lewis-Glenn researchers published hundreds of technical notes, research reports, research memoranda, and a variety of other documents on icing research. Although mainly highly technical in nature, these reports also usually contained descriptive information on the background of various projects. Several of these reports merit special mention. *Selected Bibliography of NACA-NASA Aircraft Icing Publications*, NASA Technical Memorandum 81651 (August 1981), not only surveys the literature for the 1944–1958 era of icing research, but also contains microfiche copies of all important technical documents. It is a gold mine of information. Also, John J. Reinmann and his colleagues issued a series of reports on the state of icing research at NASA. They are Reinmann, R. J. Shaw, and W. A. Olsen, Jr., *Aircraft Icing Research at NASA*, NASA Technical Memorandum 82919 (1982); Reinmann, Shaw, and Olsen, *NASA Lewis Research Center's Program on Icing Research*, NASA Technical Memorandum 83031 (1983); Reinmann, Shaw, and Richard J. Ranaudo, *NASA's Program on Icing Research and Technology*, NASA Technical Memorandum 101989 (1989); Reinmann, *NASA's Aircraft Icing Technology Program*, NASA Technical Memorandum 104518 (1991); and Reinmann, "Icing: Accretion, Detection, Protection," in Advisory Group for Aerospace Research and Development, *AGARD Lecture Series 197: Flight in an Adverse Environment* (NATO, 1994), section 4, pp. 1–27. Copies of the NACA/NASA technical documents can be found in most major research libraries. Also, a number of the reports are now available on the Worldwide Web at *http://naca.larc.nasa.gov* and *http://techreports.larc.nasa.gov*. The Glenn Research Center has a Web site at *http://www.grc.nasa.gov*. For the icing branch, see *http://icebox.grc.nasa.gov*. The IRT has its own Web site at *http://www.grc.nasa.gov/WWW/IRT/*

While the technical reports document the nature of the NACA/NASA icing research, they have to be supplemented by oral history in order to understand the context in which they were done. A list of individuals who contributed to this study follows. An * indicates that a taped interview was conducted and a copy is located in the NASA History Office in Washington, DC, along with pertinent correspondence.

Richard I. Adams
David N. Anderson*
Roger J. Aubert
Colin S. Bidwell

Bernard J. Blaha
Thomas H. Bond*
George E. Cain
Seetharam H. Chintamani
Harold H. Christenson
David Ellis
John H. Enders
Robert J. Flemming
Thomas F. Gelder*
Joseph Gerardi
Uwe von Glahn
David K. Henry
Eugene G. Hill
Thomas B. Irvine*
Susan L. Kevdzija
James K. Luers
Porter J. Perkins*
Andrew A. Peterson
Richard J. Ranaudo
Thomas P. Ratvasky*
Andrew L. Reehorst*
John J. Reinmann*
Robert J. Shaw*
Jaiwon Shin
Raymond G. Sotos*
David A. Spera
David Sweet
Allen R. Tobiason
David W. Vincent*
Halbert Whitaker
Harold Zager*
Jane Gavlak Zager*

NASA History Series

Reference Works, NASA SP-4000:

Grimwood, James M. *Project Mercury: A Chronology*. NASA SP-4001, 1963.

Grimwood, James M., and C. Barton Hacker, with Peter J. Vorzimmer. *Project Gemini Technology and Operations: A Chronology*. NASA SP-4002, 1969.

Link, Mae Mills. *Space Medicine in Project Mercury*. NASA SP-4003, 1965.

Astronautics and Aeronautics, 1963: Chronology of Science, Technology, and Policy. NASA SP-4004, 1964.

Astronautics and Aeronautics, 1964: Chronology of Science, Technology, and Policy. NASA SP-4005, 1965.

Astronautics and Aeronautics, 1965: Chronology of Science, Technology, and Policy. NASA SP-4006, 1966.

Astronautics and Aeronautics, 1966: Chronology of Science, Technology, and Policy. NASA SP-4007, 1967.

Astronautics and Aeronautics, 1967: Chronology of Science, Technology, and Policy. NASA SP-4008, 1968.

Ertel, Ivan D., and Mary Louise Morse. *The Apollo Spacecraft: A Chronology, Volume I, Through November 7, 1962*. NASA SP-4009, 1969.

Morse, Mary Louise, and Jean Kernahan Bays. *The Apollo Spacecraft: A Chronology, Volume II, November 8, 1962–September 30, 1964*. NASA SP-4009, 1973.

Brooks, Courtney G., and Ivan D. Ertel. *The Apollo Spacecraft: A Chronology, Volume III, October 1, 1964–January 20, 1966.* NASA SP-4009, 1973.

Ertel, Ivan D., and Roland W. Newkirk, with Courtney G. Brooks. *The Apollo Spacecraft: A Chronology, Volume IV, January 21, 1966–July 13, 1974.* NASA SP-4009, 1978.

Astronautics and Aeronautics, 1968: Chronology of Science, Technology, and Policy. NASA SP-4010, 1969.

Newkirk, Roland W., and Ivan D. Ertel, with Courtney G. Brooks. *Skylab: A Chronology.* NASA SP-4011, 1977.

Van Nimmen, Jane, and Leonard C. Bruno, with Robert L. Rosholt. *NASA Historical Data Book, Volume I: NASA Resources, 1958–1968.* NASA SP-4012, 1976, rep. ed. 1988.

Ezell, Linda Neuman. *NASA Historical Data Book, Volume II: Programs and Projects, 1958–1968.* NASA SP-4012, 1988.

Ezell, Linda Neuman. *NASA Historical Data Book, Volume III: Programs and Projects, 1969–1978.* NASA SP-4012, 1988.

Gawdiak, Ihor Y., with Helen Fedor, compilers. *NASA Historical Data Book, Volume IV: NASA Resources, 1969–1978.* NASA SP-4012, 1994.

Rumerman, Judy A., compiler. *NASA Historical Data Book, 1979–1988: Volume V, NASA Launch Systems, Space Transportation, Human Spaceflight, and Space Science.* NASA SP-4012, 1999.

Rumerman, Judy A., compiler. *NASA Historical Data Book, Volume VI: NASA Space Applications, Aeronautics and Space Research and Technology, Tracking and Data Acquisition/Space Operations, Commercial Programs, and Resources, 1979–1988.* NASA SP-2000-4012, 2000.

Astronautics and Aeronautics, 1969: Chronology of Science, Technology, and Policy. NASA SP-4014, 1970.

Astronautics and Aeronautics, 1970: Chronology of Science, Technology, and Policy. NASA SP-4015, 1972.

Astronautics and Aeronautics, 1971: Chronology of Science, Technology, and Policy. NASA SP-4016, 1972.

Astronautics and Aeronautics, 1972: Chronology of Science, Technology, and Policy. NASA SP-4017, 1974.

Astronautics and Aeronautics, 1973: Chronology of Science, Technology, and Policy. NASA SP-4018, 1975.

Astronautics and Aeronautics, 1974: Chronology of Science, Technology, and Policy. NASA SP-4019, 1977.

Astronautics and Aeronautics, 1975: Chronology of Science, Technology, and Policy. NASA SP-4020, 1979.

Astronautics and Aeronautics, 1976: Chronology of Science, Technology, and Policy. NASA SP-4021, 1984.

Astronautics and Aeronautics, 1977: Chronology of Science, Technology, and Policy. NASA SP-4022, 1986.

Astronautics and Aeronautics, 1978: Chronology of Science, Technology, and Policy. NASA SP-4023, 1986.

Astronautics and Aeronautics, 1979–1984: Chronology of Science, Technology, and Policy. NASA SP-4024, 1988.

Astronautics and Aeronautics, 1985: Chronology of Science, Technology, and Policy. NASA SP-4025, 1990.

Noordung, Hermann. *The Problem of Space Travel: The Rocket Motor.* Edited by Ernst Stuhlinger and J. D. Hunley, with Jennifer Garland. NASA SP-4026, 1995.

Astronautics and Aeronautics, 1986–1990: A Chronology. NASA SP-4027, 1997.

Astronautics and Aeronautics, 1990–1995: A Chronology. NASA SP-2000-4028, 2000.

Management Histories, NASA SP-4100:

Rosholt, Robert L. *An Administrative History of NASA, 1958–1963.* NASA SP-4101, 1966.

Levine, Arnold S. *Managing NASA in the Apollo Era.* NASA SP-4102, 1982.

Roland, Alex. *Model Research: The National Advisory Committee for Aeronautics, 1915–1958.* NASA SP-4103, 1985.

Fries, Sylvia D. *NASA Engineers and the Age of Apollo.* NASA SP-4104, 1992.

Glennan, T. Keith. *The Birth of NASA: The Diary of T. Keith Glennan.* J. D. Hunley, editor. NASA SP-4105, 1993.

Seamans, Robert C., Jr. *Aiming at Targets: The Autobiography of Robert C. Seamans, Jr.* NASA SP-4106, 1996.

Project Histories, NASA SP-4200:

Swenson, Loyd S., Jr., James M. Grimwood, and Charles C. Alexander. *This New Ocean: A History of Project Mercury.* NASA SP-4201, 1966; rep. ed. 1998.

Green, Constance McLaughlin, and Milton Lomask. *Vanguard: A History.* NASA SP-4202, 1970; rep. ed. Smithsonian Institution Press, 1971.

Hacker, Barton C., and James M. Grimwood. *On Shoulders of Titans: A History of Project Gemini.* NASA SP-4203, 1977.

Benson, Charles D., and William Barnaby Faherty. *Moonport: A History of Apollo Launch Facilities and Operations.* NASA SP-4204, 1978.

Brooks, Courtney G., James M. Grimwood, and Loyd S. Swenson, Jr. *Chariots for Apollo: A History of Manned Lunar Spacecraft.* NASA SP-4205, 1979.

Bilstein, Roger E. *Stages to Saturn: A Technological History of the Apollo/Saturn Launch Vehicles.* NASA SP-4206, 1980, rep. ed. 1997.

SP-4207 not published.

Compton, W. David, and Charles D. Benson. *Living and Working in Space: A History of Skylab*. NASA SP-4208, 1983.

Ezell, Edward Clinton, and Linda Neuman Ezell. *The Partnership: A History of the Apollo-Soyuz Test Project*. NASA SP-4209, 1978.

Hall, R. Cargill. *Lunar Impact: A History of Project Ranger*. NASA SP-4210, 1977.

Newell, Homer E. *Beyond the Atmosphere: Early Years of Space Science*. NASA SP-4211, 1980.

Ezell, Edward Clinton, and Linda Neuman Ezell. *On Mars: Exploration of the Red Planet, 1958–1978*. NASA SP-4212, 1984.

Pitts, John A. *The Human Factor: Biomedicine in the Manned Space Program to 1980*. NASA SP-4213, 1985.

Compton, W. David. *Where No Man Has Gone Before: A History of Apollo Lunar Exploration Missions*. NASA SP-4214, 1989.

Naugle, John E. *First Among Equals: The Selection of NASA Space Science Experiments*. NASA SP-4215, 1991.

Wallace, Lane E. *Airborne Trailblazer: Two Decades with NASA Langley's Boeing 737 Flying Laboratory*. NASA SP-4216, 1994.

Butrica, Andrew J., editor. *Beyond the Ionosphere: Fifty Years of Satellite Communication*. NASA SP-4217, 1997.

Butrica, Andrew J. *To See the Unseen: A History of Planetary Radar Astronomy*. NASA SP-4218, 1996.

Mack, Pamela E., editor. *From Engineering Science to Big Science: The NACA and NASA Collier Trophy Research Project Winners*. NASA SP-4219, 1998.

Reed, R. Dale, with Darlene Lister. *Wingless Flight: The Lifting Body Story*. NASA SP-4220, 1997.

Heppenheimer, T. A. *The Space Shuttle Decision: NASA's Search for a Reusable Space Vehicle.* NASA SP-4221, 1999.

Hunley, J. D., editor. *Toward Mach 2: The Douglas D-558 Program.* NASA SP-4222, 1999.

Swanson, Glen E., editor. *"Before this Decade is Out . . .": Personal Reflections on the Apollo Program.* NASA SP-4223, 1999.

Tomayko, James E. *Computers Take Flight: A History of NASA's Pioneering Digital Fly-by-Wire Project.* NASA SP-2000-4224, 2000.

Morgan, Clay. *Shuttle-Mir: The U.S. and Russia Share History's Highest Stage.* NASA SP-2001-4225, 2001.

Leary, William M. *"We Freeze to Please": A History of NASA's Icing Research Tunnel and the Quest for Flight Safety.* NASA SP-2002-4226, 2002.

Mudgway, Douglas J. *Uplink-Downlink: A History of the Deep Space Network 1957–1997.* NASA SP-2001-4227, 2001.

Center Histories, NASA SP-4300:

Rosenthal, Alfred. *Venture into Space: Early Years of Goddard Space Flight Center.* NASA SP-4301, 1985.

Hartman, Edwin P. *Adventures in Research: A History of Ames Research Center, 1940–1965.* NASA SP-4302, 1970.

Hallion, Richard P. *On the Frontier: Flight Research at Dryden, 1946-1981.* NASA SP-4303, 1984.

Muenger, Elizabeth A. *Searching the Horizon: A History of Ames Research Center, 1940–1976.* NASA SP-4304, 1985.

Hansen, James R. *Engineer in Charge: A History of the Langley Aeronautical Laboratory, 1917–1958.* NASA SP-4305, 1987.

Dawson, Virginia P. *Engines and Innovation: Lewis Laboratory and American Propulsion Technology.* NASA SP-4306, 1991.

Dethloff, Henry C. *"Suddenly Tomorrow Came . . .": A History of the Johnson Space Center.* NASA SP-4307, 1993.

Hansen, James R. *Spaceflight Revolution: NASA Langley Research Center from Sputnik to Apollo.* NASA SP-4308, 1995.

Wallace, Lane E. *Flights of Discovery: 50 Years at the NASA Dryden Flight Research Center.* NASA SP-4309, 1996.

Herring, Mack R. *Way Station to Space: A History of the John C. Stennis Space Center.* NASA SP-4310, 1997.

Wallace, Harold D., Jr. *Wallops Station and the Creation of the American Space Program.* NASA SP-4311, 1997.

Wallace, Lane E. *Dreams, Hopes, Realities: NASA's Goddard Space Flight Center, The First Forty Years.* NASA SP-4312, 1999.

Dunar, Andrew J., and Stephen P. Waring. *Power to Explore: A History of the Marshall Space Flight Center.* NASA SP-4313, 1999.

Bugos, Glenn E. *Atmosphere of Freedom: Sixty Years at the NASA Ames Research Center* NASA SP-2000-4314, 2000.

General Histories, NASA SP-4400:

Corliss, William R. *NASA Sounding Rockets, 1958–1968: A Historical Summary.* NASA SP-4401, 1971.

Wells, Helen T., Susan H. Whiteley, and Carrie Karegeannes. *Origins of NASA Names.* NASA SP-4402, 1976.

Anderson, Frank W., Jr. *Orders of Magnitude: A History of NACA and NASA, 1915–1980.* NASA SP-4403, 1981.

Sloop, John L. *Liquid Hydrogen as a Propulsion Fuel, 1945–1959.* NASA SP-4404, 1978.

Roland, Alex. *A Spacefaring People: Perspectives on Early Spaceflight.* NASA SP-4405, 1985.

Bilstein, Roger E. *Orders of Magnitude: A History of the NACA and NASA, 1915–1990.* NASA SP-4406, 1989.

Logsdon, John M., ed., with Linda J. Lear, Jannelle Warren-Findley, Ray A. Williamson, and Dwayne A. Day. *Exploring the Unknown: Selected Documents in the History of the U.S. Civil Space Program, Volume I, Organizing for Exploration.* NASA SP-4407, 1995.

Logsdon, John M., editor, with Dwayne A. Day and Roger D. Launius. *Exploring the Unknown: Selected Documents in the History of the U.S. Civil Space Program, Volume II, Relations with Other Organizations.* NASA SP-4407, 1996.

Logsdon, John M., ed., with Roger D. Launius, David H. Onkst, and Stephen J. Garber. *Exploring the Unknown: Selected Documents in the History of the U.S. Civil Space Program, Volume III, Using Space.* NASA SP-4407, 1998.

Logsdon, John M., gen. ed., with Ray A. Williamson, Roger D. Launius, Russell J. Acker, Stephen J. Garber, and Jonathan L. Friedman. *Exploring the Unknown: Selected Documents in the History of the U.S. Civil Space Program, Volume IV, Accessing Space.* NASA SP-4407, 1999.

Logsdon, John M., general editor, with Amy Paige Snyder, Roger D. Launius, Stephen J. Garber, and Regan Anne Newport. *Exploring the Unknown: Selected Documents in the History of the U.S. Civil Space Program, Volume V, Exploring the Cosmos.* NASA SP-2001-4407, 2001.

Siddiqi, Asif A. *Challenge to Apollo: The Soviet Union and the Space Race, 1945–1974.* NASA SP-2000-4408, 2000.

Monographs in Aerospace History, NASA SP-4500:

Launius, Roger D. and Aaron K. Gillette, comps. *Toward a History of the Space Shuttle: An Annotated Bibliography.* Monograph in Aerospace History, No. 1, 1992.

Launius, Roger D., and J. D. Hunley, comps. *An Annotated Bibliography of the Apollo Program*. Monograph in Aerospace History, No. 2, 1994.

Launius, Roger D. *Apollo: A Retrospective Analysis*. Monograph in Aerospace History, No. 3, 1994.

Hansen, James R. *Enchanted Rendezvous: John C. Houbolt and the Genesis of the Lunar-Orbit Rendezvous Concept*. Monograph in Aerospace History, No. 4, 1995.

Gorn, Michael H. *Hugh L. Dryden's Career in Aviation and Space*. Monograph in Aerospace History, No. 5, 1996.

Powers, Sheryll Goecke. *Women in Flight Research at NASA Dryden Flight Research Center From 1946 to 1995*. Monograph in Aerospace History, No. 6, 1997.

Portree, David S. F. and Robert C. Trevino. *Walking to Olympus: An EVA Chronology*. Monograph in Aerospace History, No. 7, 1997.

Logsdon, John M., moderator. *Legislative Origins of the National Aeronautics and Space Act of 1958: Proceedings of an Oral History Workshop*. Monograph in Aerospace History, No. 8, 1998.

Rumerman, Judy A., comp. *U.S. Human Spaceflight, A Record of Achievement 1961–1998*. Monograph in Aerospace History, No. 9, 1998.

Portree, David S. F. *NASA's Origins and the Dawn of the Space Age*. Monograph in Aerospace History, No. 10, 1998.

Logsdon, John M. *Together in Orbit: The Origins of International Cooperation in the Space Station*. Monograph in Aerospace History, No. 11, 1998.

Phillips, W. Hewitt. *Journey in Aeronautical Research: A Career at NASA Langley Research Center*. Monograph in Aerospace History, No. 12, 1998.

Braslow, Albert L. *A History of Suction-Type Laminar-Flow Control with Emphasis on Flight Research*. Monograph in Aerospace History, No. 13, 1999.

Logsdon, John M., moderator. *Managing the Moon Program: Lessons Learned From Apollo*. Monograph in Aerospace History, No. 14, 1999.

Perminov, V. G. *The Difficult Road to Mars: A Brief History of Mars Exploration in the Soviet Union* is Monograph in Aerospace History, No. 15, 1999.

Tucker, Tom. *Touchdown: The Development of Propulsion Controlled Aircraft at NASA Dryden.* Monograph in Aerospace History, No. 16, 1999.

Maisel, Martin D., Demo J. Giulianetti, and Daniel C. Dugan. *The History of the XV-15 Tilt Rotor Research Aircraft: From Concept to Flight.* NASA SP-2000-4517, 2000.

Jenkins, Dennis R. *Hypersonics Before the Shuttle: A Concise History of the X-15 Research Airplane.* NASA SP-2000-4518, 2000.

Chambers, Joseph R. *Partners in Freedom: Contributions of the Langley Research Center to U.S. Military Aircraft in the 1990s.* NASA SP-2000-4519, 2000.

Waltman, Gene L. *Black Magic and Gremlins: Analog Flight Simulations at NASA's Flight Research Center.* NASA SP-2000-4520, 2000.

Portree, David S. F. *Humans to Mars: Fifty Years of Mission Planning, 1950–2000.* NASA SP-2001-4521, 2001.

Thompson, Milton O., with J. D. Hunley. *Flight Research: Problems Encountered and What They Should Teach Us.* NASA SP-2000-4522, 2000.

Tucker, Tom. *The Eclipse Project.* NASA SP-2000-4523, 2000.

Index

Deemed a historic landmark by the American Society of Mechanical Engineers, the IRT is also a unique structure in and of itself. Willis Carrier, the "father of air conditioning," was a lead consultant during IRT construction in the early 1940s and considered it his company's greatest engineering achievement. The IRT's requirements for a wind tunnel and a very large refrigerated facility included a specially designed heat exchanger and spray system.

The *"We Freeze to Please"* motto also makes a special statement about the IRT and its staff. An early and major player in cooperative research with other government agencies, commercial firms, and foreign researchers, the IRT has a deserved reputation for excellent customer service. In a difficult research discipline, a flexible attitude has enabled the IRT staff to push the boundaries of their work even further.

This is the story of a unique facility that has made unparalleled contributions to a specialized area of aeronautics research that affects virtually all who fly. Operating for over half a century, the work done at the IRT continues to push the edge of the envelope, and researchers there continue to make major contributions toward a problem that plagues aircraft around the world. *"We Freeze to Please"* brings this record forward clearly for the attention of specialists, policymakers, students, and general readers.

About the Author:

Dr. William M. Leary is Coulter Professor of History at the University of Georgia, Athens. He has written numerous articles and essays on aerospace history that have appeared in scholarly journals and reference works published in America and overseas. Also, he has written or edited a number of books including *Project Coldfeet: Secret Mission to a Soviet Ice Station* (Naval Institute Press, 1996); *Codename Mule: Fighting the Secret War in Laos for the CIA* (Naval Institute Press, 1995); *Aerial Pioneers: The U.S. Air Mail Service* (Smithsonian Institution Press, 1985); *Aviation's Golden Age: Portraits from the 1920s and 1930s* (University of Iowa Press, 1989); and *The Airline Industry, a volume in the Encyclopedia of American Business History and Biography* (Facts on File, 1990).

About the Cover:

"High-lift Icing Test," oil by Linda Draper.

This painting portrays the teamwork and coordination of the test crew in the Icing Research Tunnel (IRT) at Glenn Research Center. The crew prepares the model for the next phase of testing. The NASA/Douglas High-lift Icing Test was conducted in the IRT from 19 July to 13 August 1993.